'An engaging perspective on e
tour of neuroscience efforts t
systems, provides a discussior
the human self, and gives practical tips and ways to help improve emotional self-regulation.'

— **Kent Berridge**, *James Olds Distinguished University Professor of Psychology and Neuroscience, University of Michigan, USA*

'João Ascenso provides a helpful and exhaustive mapping of the neural bases of emotions as a means for understanding the neural bases of emotional intelligence. This mapping paves the way for a useful description of emotional intelligence approaches in schools and organizations. The book serves as a valuable reference for anyone interested in the neuroscientific foundations of emotional intelligence.'

— **Jane Dutton**, *Robert L. Kahn Distinguished University Professor Emeritus of Business Administration and Psychology, University of Michigan, Ann Arbor, USA. CoDirector, Center for Positive Organizational Scholarship, and Former President, National Academy of Management*

Dear Jane,

It is with immense gratitude that I write you this message. Thank you for your wonderful experience and contribution to the field of Positive Organizational Scholarship! You have been an inspiration to me!

All the best, João

The Neuroscience of Emotional Intelligence and Its Applications to Education and Organizations

This book provides a clear understanding of the neuromechanisms of emotional intelligence and its applications to education and organizations through practical exercises.

Divided into three parts, the book begins by explaining the data that help us understand the neural mechanisms of emotional intelligence. Part 2 focuses on application in educational contexts by presenting emotional intelligence education programs for children and adolescents as well as an analysis of emotional intelligence from a practical point of view. Part 3 switches the focus to organizations through the leadership with emotional and social intelligence model as proposed and validated by Daniel Goleman and Richard Boyatzis. Both parts offer a series of practical and engaging exercises for application to adolescents, children and educators, and organizational environments, respectively. Presented simply, the book gives a scientifically rigorous and structured overview of how neuroscience has helped in understanding the neural mechanisms of emotions and its applications.

It is indispensable reading for neuroscientists, psychologists, leaders, managers, teachers, and educators, and all those interested in the search for personal and professional success.

João Ascenso is CEO of a multinational company headquartered in Dubai called International Institute for Executive Education and Positive Change for Business Prosperity L.L.C-FZ and is Senior Consultant in Organizational Behavior for Duke Kunshan University, and Future Director, Center for the Sustainable Development of Africa.

The Neuroscience of Emotional Intelligence and Its Applications to Education and Organizations

João Ascenso

LONDON AND NEW YORK

Cover Image: © Imaginima via Getty Images

First published in English 2025
by Routledge
4 Park Square, Milton Park, Abingdon, Oxon OX14 4RN

and by Routledge
605 Third Avenue, New York, NY 10158

Routledge is an imprint of the Taylor & Francis Group, an informa business

Copyright © 2025 property of FCA PACTOR Editores, Lda. all rights reserved

The right of João Ascenso to be identified as author of this work has been asserted in accordance with sections 77 and 78 of the Copyright, Designs and Patents Act 1988.

All rights reserved. No part of this book may be reprinted or reproduced or utilised in any form or by any electronic, mechanical, or other means, now known or hereafter invented, including photocopying and recording, or in any information storage or retrieval system, without permission in writing from the publishers.

Trademark notice: Product or corporate names may be trademarks or registered trademarks, and are used only for identification and explanation without intent to infringe.

Published in Portuguese by FCA PACTOR Editores, Lda. 2022

British Library Cataloguing-in-Publication Data
A catalogue record for this book is available from the British Library

ISBN: 978-1-032-82292-1 (hbk)
ISBN: 978-1-032-82291-4 (pbk)
ISBN: 978-1-003-50388-0 (ebk)

DOI: 10.4324/9781003503880

Typeset in Optima
by Apex CoVantage, LLC

Contents

Foreword	*ix*
Introduction	*xiii*

PART I
What Neuroscience Can Teach Us About Emotional Intelligence   1

1	The Neural Bases of Emotions	3
2	The Broaden-and-Build Theory of Positive Emotions	57
3	Complexity, Evolution and the Neural Bases of Self	74
4	The Brain Mechanisms of Emotional Self-Awareness	80
5	The Neural Bases of Emotional Self-Control	88
6	The Science of Human Relationships and Its Relationship With Neuroscience	99

PART II
Emotional Intelligence Applied to Education   113

7	Educational Programs in Emotional Intelligence	115
8	Dimensions of Focus of Emotional Intelligence: Self and Others	125
9	Emotional Intelligence Exercises	142

PART III
Emotional Intelligence Applied to Organizations   165

10 Leadership With Emotional and Social Intelligence   167
11 Leadership With Emotional Intelligence Exercises   191
12 Answer Suggestions for the Leadership With Emotional
    Intelligence Exercises   211

Conclusion   *226*
Index   *229*

Foreword

In this foreword, Professor Richard Boyatzis mentions the prevalence of the need for emotional intelligence throughout human evolution. He mentions the existence of two main neural networks that tend to suppress each other in practical life: the analytic network (AN) and the empathic network (EN) (i.e., technically often called the default mode network). Neuroscientific evidence suggests that they tend to suppress each other in a given situation but being emotionally intelligent means having the capacity to activate one instead of the other in an intelligent manner in practical situations, in order to be effective. In this sense, emotional intelligence means, according to Boyatzis, the capacity to interplay between these two important networks in organizational, social and personal lives.

Our ability to work and live with others has distinguished the successful adaptation of our species since discovering fire and sitting around it for warmth or cooking, and probably earlier. Scanning popular books, movies and television series suggests that we are still puzzled by it. The prevalence of books on leadership, teamwork, relationships and self-renewal speaks to the deeply felt need we all have for a sense of purpose in life as well as the desire to be with others. Although articulated as key capabilities since the writings of Aristotle and Confucius, it has been only in the last 30 years that we have acknowledged these capabilities and called them emotional intelligence.

The ability to manage our own emotions (i.e., emotional intelligence) as well as work with the emotions of others to build better relationships (i.e., social intelligence) are key to effectiveness at work, having a caring family, and contributing to society. And yet scholars still argue about its definitions, measurement and relevance in life and work. Why? The essence of the basis for such confusion lay within the neural networks and deeply unconscious functioning of our brain. Then it all emerges as our behavior.

Emotional and social intelligence (ESI) exist within a person at multiple levels. We each have some degree of dispositional characteristics, like performance ability or traits of ESI, except for those on the autism spectrum. We also each have some aspect of our self-image that includes ESI. The most evident level of ESI emerges in our behavior and how others experience us. This behavioral level of ESI is the one most related to work effectiveness and how well we get along with family and friends.

Some confusion in the field is caused by scholars arguing about theory or measures, when, in fact, all are correct but with different implications. Dispositional ESI (i.e., trait or ability) is often statistically related to analytic intelligence (i.e., traditional general mental ability). Other trait measures, such as self-perception or self-assessed measures, are often statistically related to personality traits. The behavioral measures have been found to have predictive statistical power to effectiveness, engagement, organizational citizenship, and innovation, controlling for general mental ability and personality. When we describe someone as showing or even having more ESI, we often are commenting on their behavior as we and others have observed it.

Why do some people in the same fields and jobs, whether executives, professionals, physicians, nurses, professors, or dentists, show consistently more ESI than others? Given how strongly ESI is related to effectiveness in management and professions, it is a puzzle as to why people would continue to act as if they do not even know how to spell ESI.

Differences become clearer once we examine internal psycho-physiological processes, especially through neuroimaging and hormonal studies. In the last 25 years, neuroscience has leaped beyond the simplistic notions of left versus right brain orientation or claims to train people through neurolinguistics programming or other such pop-psychology fads. For example, we know that among other important networks, we have two neural networks that suppress each other. As Professor Anthony Jack of Case Western Reserve University has repeatedly shown, the AN (i.e., technically often called the task positive network) activates in opposition to the EN (i.e., technically often called the default mode network). The AN helps us solve problems, make decisions, focus on tasks, and do anything involving analytics from philosophy to financial analysis. Meanwhile, the EN enables us to be open to new ideas, scan the environment for trends and changes, and be sensitive to other people and emotions.

Again, these two critical networks *suppress* each other. No wonder executives preoccupied and focused on their financial performance have difficulty motivating and engaging their staff. Since it appears that these neural networks can be developed and enhanced with practice, we have hope that advances in neuroscience can help managers, professors and teachers and those in helping professions to develop facility with both networks and an ease of knowing which to activate when.

Activation of these neural networks leads to ESI dispositional traits as well as frequency of ESI behavior. As a few studies have confirmed, we need all forms of intelligence to be effective in management and professional roles. We need cognitive, emotional, and social intelligence. No one form of intelligence can compensate for an absence of or difficulty using the other forms.

Since these neural networks activate and function at amazing speeds, the processes are unconscious. The results are not. They are observed by anyone around us.

Developing our ESI and becoming more effective in leadership and management, professions and even being a family member or friend, involve using conscious choices to alter our behavior. Research has given us hope in showing that adults can develop ESI. Practicing new ESI behavior, thoughts, feelings, or attitudes can result in sustainable new habits and patterns of behavior.

As a science in its early stages, understanding neural processing, including the sympathetic and parasympathetic nervous systems (i.e., the human stress and renewal hormonal systems), is key to understanding ESI and desired change. Our challenge is to stay current. This is where works like this book by João Ascenso become important allies in our own development. This is not a call for each of us to become neuroscientists or revel in being nerdy. If the reader is new to the fields of emotional intelligence or social neuroscience, I recommend you read the book in small doses. You might even jump around and read sections or chapters that are of immediate interest first and then backtrack to earlier chapters.

For the accomplished reader, have all of it! With either approach, you will gain insight into how and why your neural systems are activated and affect your body, mind, mood, and behavior. It will help you understand how ESI emerges in how we act and helps or hinders our relationships.

Richard Boyatzis, PhD (Harvard University), is Distinguished University Professor, Case Western Reserve University, as well as Professor, Departments of Organizational Behavior, Psychology and Cognitive Science. He is co-author of the international bestseller *Primal Leadership* and the book *Helping People Change*.

Introduction

This introduction aims to give the reader the notion that it was the scientific contribution of Professor Antonio Damasio that prompted the outbreak of many laboratories of psychology, on the one hand, and of neuroscience, on the other, by creating entire avenues of scientific research dedicated to the study of emotions and their relationships with specific mental processes and social and moral behaviors.

Of particular interest to this book, Professor Damasio was one of the most famous scientists who contributed to the creation of the concept of "emotional intelligence." When Damasio began experimentation on patients with brain lesions in the 1980s to understand the relationships between emotions and thought processes and with social and moral behavior, the concept of "emotional intelligence," from a scientific point of view, had not yet been created.

This concept was created only in 1990 by the researchers Peter Salovey and John Mayer, 5 years after the publication of their article in the scientific journal *Neurology* (Eslinger & Damasio, 1985; Salovey & Mayer, 1990). The first scientific study with patients with brain lesions drew the world's attention to the fact that emotions are fundamental not only in our social life but also in our intrapersonal world, such as in our decision-making and thought processes.

I also take this opportunity to express my gratitude to Professor Richard Boyatzis, whose pioneering contribution to scientific research into competencies in organizations and the scientific validation of Daniel Goleman's emotional intelligence competency model shaped my entire career as a trainer in leadership with emotional intelligence since 2005 and gave scientific dignity to emotional intelligence applied to organizations. Daniel Goleman and his ideas and applications to organizations have transformed my career in executive education from the beginning.

I also thank him for his kindness for giving me the honor of writing the preface of this book.

This book is structured in three parts.

In Part I, "What Neuroscience Can Teach Us About Emotional Intelligence," the focus is to transmit in a simple way to the reader neuroscience data that contribute to a better understanding of the neural mechanisms of emotional intelligence.

We start with a definition and discussion of the psychological functions of all basic and moral emotions and their neural basis with the aim of helping the reader increase their emotional self-awareness in some of the emotions mentioned through its scientific description. After that, we present a comprehensive description of all the positive emotions studied scientifically, as well as the important broaden-and-build theory of positive emotions and its applications in daily life. Next, we give a notion of the neural basis of the self and a notion of the life goal of everyone, which is evolution. This part explains what the process of evolution of the self consists of. Moreover, this part also describes in detail the brain mechanisms of emotional self-awareness and emotional self-control, and the relationship of the science of human relationships with neuroscience.

In Part II, "Emotional Intelligence Applied to Education," the purpose is to present emotional intelligence education programs for children and adolescents, called emotional and social learning, and the two levels of analysis of emotional intelligence from a practical point of view – "self" and "others" – and finally provide practical emotional intelligence exercises for children and adolescents as well as for educators.

In Part III, "Emotional Intelligence Applied to Organizations," the main purpose is to explain in detail the applications of emotional intelligence to organizations, through the leadership with emotional and social intelligence model proposed by Daniel Goleman and scientifically validated by Professor Richard Boyatzis, as well as the explanation of each model's competence, its evolution and leadership styles associated with the model. We also intend in this part to offer the reader practical leadership exercises, with the aim of helping the reader to practice the leadership styles and competencies of leadership with emotional and social intelligence in practical situations.

I sincerely hope that this book will be useful in some way to readers, especially in their practical, personal, relational, or professional lives.

References

Eslinger, P. J., & Damasio, A. R. (1985). Severe disturbance of higher cognition after bilateral frontal lobe ablation: Patient EVR. *Neurology, 35*(12), 1731–1731. https://doi.org/10.1212/WNL.35.12.1731

Salovey, P., & Mayer, J. D. (1990). Emotional intelligence. *Imagination, Cognition, and Personality, 9*(3), 185–211. https://doi.org/10.2190/DUGG-P24E-52WK-6CDG

Part I

What Neuroscience Can Teach Us About Emotional Intelligence

Chapter 1

The Neural Bases of Emotions

The Neural Bases of Emotions

Before presenting the neural bases of emotions, it is important to understand the main emotions that exist in human beings, which ones we share with other animals, which are exclusive of human beings, their scientific definitions, their characteristic mechanisms, their specific behavioral tendencies, and the specificity of psychological functions from an evolutionary point of view.

From a scientific point of view, emotions are divided into basic emotions (Ekman, 1993) and moral emotions (Haidt, 2003a). The basic emotions are those that manifest themselves universally, while moral emotions have cultural differences in their expression (Haidt, 2003a). In addition to basic and moral emotions, we also describe in detail the main positive emotions studied scientifically in the last 30 years and their neural bases, although there are still many positive emotions that have not been studied in social neuroscience, simply because this is an area of research that is still in its infancy. For many decades there has been an excess of study of negative emotions and a lack of study of positive emotions.

Fortunately, in the last 20 years, there was a significant increase in scientific studies of positive emotions (Fredrickson, 2013).

Basic Emotions

There are two fundamental characteristics that allow you to designate an emotion as basic. The first is the fact that the basic emotions are discrete, that is, they can be fundamentally distinguished from each other (Ekman & Cordaro, 2011). Scientific data from discrete emotions include a universal physiological expression that is invariable from culture to culture, namely, facial and vocal physiology, a unique autonomic physiology, and a study of the events preceding the emotion (Ekman & Davidson, 1994).

The second feature is that basic emotions evolve through adaptation to the surrounding environments to which humans are subjected (Ekman

DOI: 10.4324/9781003503880-2

& Cordaro, 2011). Thus, basic emotions represent biological mechanisms that enable human beings to react to fundamental tasks in life, such as survival, losses, successes and joys. Each basic emotion stimulates specific action tendencies, and these actions contribute decisively to our survival and evolution, and are, from the evolutionary point of view, much more effective in practical solutions when compared with other solutions to achieve species' goals. In addition, the basic emotions are those that we share with other animals (Darwin, 1965).

Another important aspect is that emotions can be expressed constructively or destructively (Lama & Ekman, 2008). And this is what constitutes the essence of emotional intelligence: deal with emotions with intelligence, to use the emotions constructively, both in the relationship with oneself and in relationships with others.

There is consensus among leading scientists as to the existence of at least seven basic emotions: fear, anger, sadness, disgust, surprise, contempt, and happiness (or joy) (Ekman & Cordaro, 2011).

Although contempt is considered a basic emotion, it is described in the moral emotions group (Haidt, 2003a).

Fear

Fear is an emotional response to the threat of harm, whether physical or psychological. Fear activates behavioral impulses to paralyze or escape. Usually fear tends to trigger anger (Ekman & Cordaro, 2011). The basic emotions are biological systems that lead animals and humans to reach the survival objective (Darwin, 1965). The emotion of fear is one of the most important basic emotions because its main objective is to guarantee self-preservation through the struggle for survival. Fear is the most scientifically studied emotion by both psychologists and neuroscientists.

The first reference on the physical manifestations of fear can be found in the *Principles of Psychology*, of Herbert Spencer (Spencer, 1855): "Fear, when felt intensely, expresses itself in cries, in efforts to hide or escape, in palpitations and tremors; and those are only the manifestations that would accompany a real experience of fear" (Spencer, 1855, pp. 356–357).

On fear, Charles Darwin expresses himself as follows:

> 'Fear, once again, is the most depressing of all emotions: and soon inducing total and forsaude, as if, as a result of or in association with the most violent and prolonged attempts to escape from danger, though none of these attempts were actually made. However, even extreme fear usually acts at first as a powerful stimulant. A man or animal taken from terror to despair is endowed with wonderful strength and is notoriously dangerous to the highest degree.'
>
> (Darwin, 1965, p. 81)

In this quote from Darwin (1965), it is clear the emotion of fear can play a key role in the struggle for survival, or in the induction into action to escape, or in the force that can be provided by turning it into anger.

Emotionally intelligent (constructive) management of fear means the capacity to transform fear into prudence and attention to detail, to anticipate obstacles and to develop strategies capable of circumventing or overcoming them intelligently.

Non-intelligent (destructive) management of fear means that the person allows fear to lead to behavioral paralysis or passivity, or trigger uncontrolled anger, or stop one from looking for one's own goals, or allow it to provoke destructive thoughts (cognitions) or inhibitors of personal development.

The Neural Bases of Fear

A study by Professor Antonio Damasio at the University of Iowa on patients with brain injuries showed that fear is associated with the amygdala (Figure 1.1) (Adolphs et al., 1994). Through the study of a patient with a rare lesion in the amygdala, the teacher's team identified that this patient was unable to detect facial expressions of fear, despite being able to identify the visualized faces (Adolphs et al., 1994).

A sophisticated statistical analysis of 82 neuroscience studies confirmed the idea that fear is associated with the bilateral amygdala, the right cerebellum and the right insula, with greater predominance for the former (Vytal & Hamann, 2010).

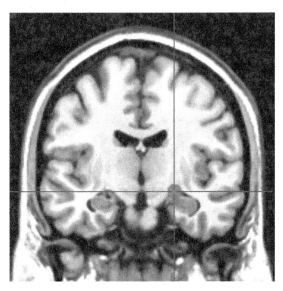

Figure 1.1 Representation of the right amygdala, according to an image of functional magnetic resonance imaging, at the intersection of the horizontal line with the vertical line.

Anger

Anger is the emotional response to negative interference over a goal that is important for the self. It can be triggered by someone who causes harm (physically or psychologically) or someone held in our esteem. Anger involves the tendency to remove the obstacle or to stop the damage, and, at its limit, involves the desire to harm the cause of interference (Ekman & Cordaro, 2011).

According to Charles Darwin, anger is the emotion that most quickly induces action and that is externalized in physical expressions of both the body and the face (Darwin, 1965).

Note a reference by Charles Darwin to the emotional reaction of a mother who watches her child be injured:

> 'But let anyone intentionally hurt your child and see the transformation! How does it start with a menacing look, such as your eyes shine and her face turns red, as her breasts shake, his narins dilate and his heart beats; for anger, not love has usually led to action.'
>
> (Darwin, 1965, p. 78)

It is precisely this rapid tendency toward action, wired in a biological system, that has allowed anger to ensure the survival of species (Darwin, 1965).

Emotionally intelligent (constructive) management of anger transforms it into self-affirmation, self-value and self-virtues in reaction to an aggression or depreciation from others. Without going to aggression, anger is turned into determination to achieve a goal.

Emotional intelligence means using the emotional impulse of anger to transform it into a constructive action, that is, its management is done in the right way, for the right reason, at the right time and with the right person (Aristotle, 2011). When anger is transformed into constructive action, the emotional content of this emotion is automatically reconfigured positively, which causes anger to diminish its emotional and biological manifestation and its original tendency to desire destruction or aggression.

For emotionally intelligent management of anger, a cognitive reappraisal is also important to transform negative thoughts regarding the target of anger. If this does not happen, every time the target of anger appears in memory, the emotion of anger will be triggered. On the one hand, it is not appropriate to feed (in one's own mind) these negative thoughts to expand mental stimuli to the emotion of anger. On the other hand, a cognitive reappraisal is to be performed toward the target of anger, to, in the long run, transform the emotion of anger directed at it. First, it is suggested that the emotional and behavioral aspects of anger be reduced, and then one may proceed to a cognitive reappraisal, so that in the future, the target (the stimulus) does not automatically trigger anger. It is very important for the emotional management of anger that it does not become a chronic emotion in the individual (Goleman, 2004).

A way to manage anger destructively is to transform, in behavioral terms, anger into uncontrolled aggression and disruptive behaviors, leading to desires and plans for revenge, in a professional or personal life, disguised or explicit. This may then turn it into hatred, or lead to arguments and relationship conflicts that are offensive and lead to relationship breakups. For example, it is known that more than 50 percent of couples during the first 7 years of marriage tend to break up due to arguments and conflicts characterized by the emotion of anger (Gottman & Levenson, 2000). Couples who learn to manage the emotion of anger during discussions and conflicts tend to survive the first 7 years of marriage (Gottman & Levenson, 2000).

The Neural Bases of Anger

A study that identified the amygdala associated with anger using a questionnaire that assessed anger as an emotional trait verified a correlation with the activation of the amygdala in a resting state (Fulwiler et al., 2012). In addition to associating the amygdala with the emotional trait of anger, the researchers noted that individuals characterized by this trait tend to have an inverse coupling between the activation of the amygdala and the orbitofrontal cortex, i.e., when in a state of mental rest state, the amygdala tends to be activated and the orbitofrontal cortex deactivated.

This study suggests that people who have anger as an emotional trait have activation of the amygdala (Figure 1.2) and deactivation of the orbitofrontal

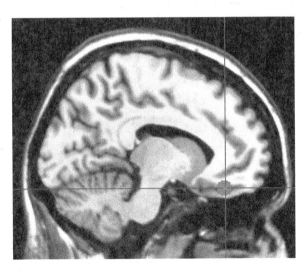

Figure 1.2 Representation of the orbitofrontal cortex, according to functional magnetic resonance imaging, at the intersection of the horizontal line with the vertical line.

cortex (Fulwiler et al., 2012). Thus, the greater the anger as an emotional trait, the lower the relationship between the activation of the amygdala and the orbitofrontal cortex. On the other hand, the attempt at self-control of anger increased the relationship of the orbitofrontal cortex with the amygdala in a resting state (Fulwiler et al., 2012). These data suggest that the orbitofrontal cortex is the region responsible for emotional self-control of anger, through the deactivation of the amygdala (Fulwiler et al., 2012).

Sadness

Sadness is the emotional response to the loss of an object or person who is very important. The typical experience of sadness is the death of a child, parent, or spouse. Usually, in sadness there is resignation, but it can become anguish, as a form of rebellion or revolt after the loss, and then a return to sadness (Ekman & Cordaro, 2011).

Sadness has a crucial adaptive function, which is to promote a deeper reflection following the loss of a person or an object that is important to the self (Lazarus, 1991). Also, it may arise after an action that the individual has taken and has subsequently acknowledged that has not been correct from a moral point of view, which follows repentance or guilt. Although guilt is characterized as anger directed at the self, it is possible that, after the anger directed toward the self, the sadness related to the self may follow. Sadness also tends to follow the emotion of repentance, and in both cases, it has the function of deepening reflection toward the self-behavior. Who has not already had the experience of reacting impulsively or inappropriately to a spouse, friend, or child, and then felt temporarily sad about yourself for it? This is a very common event in human relationships.

The experience of sadness turns the attention of the human being to the self, promoting deeper reflections about the behavior of the self and others; deepens the reflection on values, personal priorities and life goals; and tends to promote resignation in the face of the acceptance of events, self-acceptance, or acceptance of others (Izard, 1977, 1993).

During the experience of sadness, physiological activation tends to decrease, allowing more "time" to spend on self-directed reflection, so that cognitive processes of assimilation and accommodation of values occur within the self.

Sadness also enables the self-assessment of our values and life plans (Bonanno & Keltner, 1997). Probably, if we did not feel sadness, but only joy, we would not develop the ability to reflect on ourselves, on others and on the values that are most important to us.

We would be more superficial, with weaker self-awareness and with lower awareness of others. One of the risks of frequently experiencing the emotion of sadness is that it could lead to depression. Sadness needs to be appreciated in its cognitive function while deepening relevant information on the values of life and of the people important to each one of us, since it is a facilitating element of a moral assessment of one's own behavior. However, it needs to be alternated with the thrill of joy to avoid the occurrence of depression.

An example of the constructive function of the emotion of sadness is what researchers designate as post-traumatic growth (Calhoun & Tedeschi, 2014; Tedeschi & Calhoun, 2004), which is a set of personal growth and positive changes that occurs in people after suffering some type of trauma, characterized by a greater appreciation of life in general, interpersonal relationships with a deeper meaning, a greater sense of inner strength, a change in life priorities based on intrinsic developmental personal values, an increase in the desire to help others and form relationships with a deeper meaning and a richer existential orientation (Tedeschi & Calhoun, 2004).

Although these researchers do not explicitly mention the role of the emotion of sadness as a mediator of the post-traumatic growth process, it is assumed that this emotion plays an important role in the emotional acceptance of a traumatic event without revolt (rather than generating anger or revolt, which would be a sign of non-acceptance of the event), and its use for the psychological growth of the self, and this influence in all spheres of a person's intra- and interpersonal life.

In the future, scientific studies will be needed to test this hypothesis. Scientific data show that when people are under the influence of sadness, they tend to perform more systematic and detailed processing of information and to reflect more on the self and its core values of life – called central values for the self, or self-centrality (Verplanken & Holland, 2002; Schwarz, 1998).

In addition, people under the effect of sadness tend to resort less to shortcuts to make fewer decision-making errors based on heuristics (Kahneman, 2011), to rely less on first impressions and analyze in more detail the available information (Schwarz, 1990). So, it is clear that sadness can lead to deeper and more accurate reflections in comparison, for example, with joy (Storbeck & Clore, 2005; Schwarz, 1990). Joy generates other benefits and is described in detail in the part dedicated to positive emotions.

The Neural Basis of Sadness

At least two meta-analyses, i.e., statistical summaries of dozens of neuroscientific studies, associate the emotion of sadness with the medial prefrontal cortex (MPFC) (Figure 1.3) (Vytal & Hamann, 2010; Phan et al., 2002).

Disgust

Disgust is an emotional response of revulsion through vision, smell, or taste – that is, it is a rejection response to some physical stimulus. It differs from moral disgust, which is explained in the group of moral emotions.

Some researchers proposed that the evolutionary origins of the emotion of disgust have the objective of protecting the organism from infections

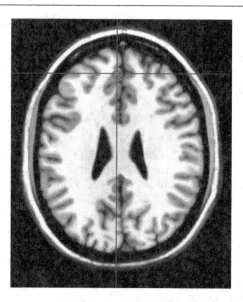

Figure 1.3 Representation of the medial prefrontal cortex, according to functional magnetic resonance imaging, at the intersection of the horizontal line with the vertical line.

(Curtis & Biran, 2001) and preserving health, both in animals and in humans. This hypothesis was confirmed by results of a large-scale international scientific study in which it was shown that disgust is a universally generated emotion in response to contact with potential stimulations of diseases, such as secretions from the body, viscous substances and sick or dirty people (Curtis et al., 2004).

Another study conducted with odors confirms similar results (Stevenson & Repacholi, 2005).

Thus, all organisms must protect themselves not only from external predators but also from microorganisms and parasites in their own bodies as well as toxic substances.

By disgust, the biological system of aversion to toxic materials or infection before their ingestion served as a useful adaptive function to survival through disease prevention.

The Neural Bases of Disgust

A meta-analysis composed of the statistical study of 83 neuroscientific studies demonstrated that the emotion of disgust is associated with the insula and orbitofrontal cortex (Figure 1.4) (Vytal & Hamann, 2010).

Another meta-analysis composed of the statistical analysis of 55 neuroscientific studies revealed that the emotion of disgust is also associated with activation of the basal ganglia (Sprengelmeyer et al., 1998). Some researchers

The Neural Bases of Emotions 11

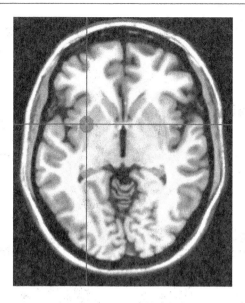

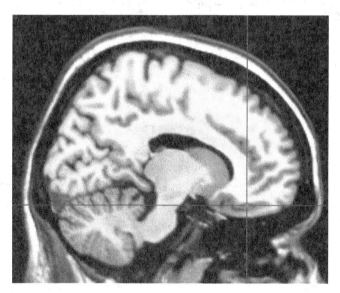

Figure 1.4 Representation of (a) the right anterior insula and (b) the orbitofrontal cortex, according to functional magnetic resonance imaging, at the intersection of the horizontal line with the vertical line.

have even proposed that the core regions of the basal ganglia are responsible for the emotion-specific processing of disgust (Sprengelmeyer et al., 1998), because patients with lesions in these regions are incapable of identifying the emotion of disgust in facial expressions (Sprengelmeyer et al., 1998).

12 What Neuroscience Can Teach Us About Emotional Intelligence

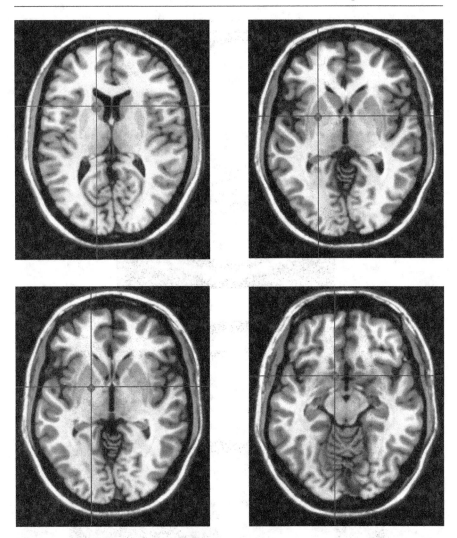

Figure 1.5 Representation of the right caudate nucleus (a), the right putamen (b), the right pallidum (c) and the nucleus accumbens (d), belonging to the basal ganglia and specific for the processing of the emotion of disgust, according to functional magnetic resonance imaging, at the intersection of the horizontal line with the vertical line.

The basal ganglia are composed of the right caudate nucleus, (Figure 1.5a), the right putamen (Figure 1.5b), the right pale globe (globo palidum) (Figure 1.5c), and the right nucleus accumbens (NAcc) (Figure 1.5d).

Surprise

Surprise is the emotional response to an unexpected event. Of all the emotions, it is the one with the shortest duration. In humans, surprise is an emotion that arises initially during the first 6 months of life. Children show surprise when events occur that they did not expect.

For example, when they see an adult who does not walk toward you, children manifest interest and surprise rather than fear or joy (Lewis, 1975). Surprise can be seen when there is no realization of an expectation or as a response to a discovery, a kind of emotional experience of "Ah!".

Surprise seems to be an emotion that allows children and adults to develop an intrinsic interest in the unknown.

The Neural Bases of Surprise

Research found a relationship between the surprise emotion and a distributed brain network, which encompasses the medial anterior cingulate cortex, the dorsal medial cingulate cortex, the supplementary premotor area, the bilateral dorsal striated body, the bilateral anterior insula, the MPFC, and the midbrain (Fouragnan et al., 2018).

Figure 1.6 presents the main brain regions associated with the emotion of surprise, according to functional magnetic resonance imaging (fMRI), at the intersection of the horizontal line with the vertical line.

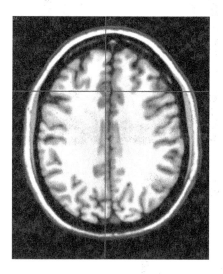
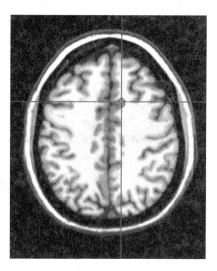

Figure 1.6 Brain regions associated with the emotion of surprise, according to functional magnetic resonance imaging, at the intersection of the horizontal line with the vertical line: (a) anterior medial cingulate cortex, (b) dorsal medial cingulate cortex, (c) right dorsal striatum, (d) left dorsal striatum, (e) right anterior insula, (f) left anterior insula, and (g) midbrain.

Happiness (Joy)

Happiness and joy are emotions appreciated and sought by people. There is not as much scientific evidence about these emotions as the previously

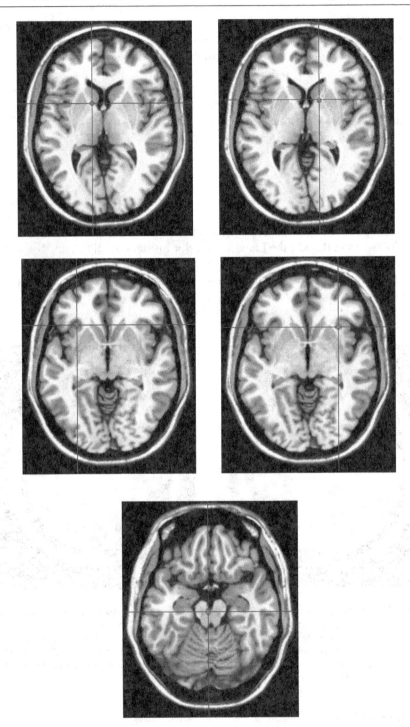

Figure 1.6 (Continued)

described negative emotions, such as fear, anger, and disgust (Ekman & Cordaro, 2011), probably because negative emotions have neural and behavioral mechanisms with different evolutionary functions and action-tendency goals than those found with positive emotions.

Joy can be considered a basic positive emotion, but happiness is a much more complex emotion. There is a whole line of scientific research pioneered by scientist and professor Ed Diener (Diener & Tay, 2017) and, later, the Nobel Prize winner in Economics, Professor Daniel Kahneman, who led a fairly large number of researchers in this line of research about happiness (Kahneman, 1999).

Happiness is a more complex concept because it encompasses cognitive components, for example, life satisfaction (Ehrhardt et al., 2000), emotional well-being (Frijda, 1999; Kahneman, 1999), economic aspects of life (Frey & Stutzer, 2018) and intentional behavioral activities that promote happiness (Lyubomirsky, 2008).

Usually, the emotion of joy is confused with euphoria and even mania, not only among the lay public but also among many researchers, both psychologists and neuroscientists.

Overly intense and positive mood, and excessive and unrealistic optimism, can generate bad decisions in business and investments as well as donations of goods without any criteria and reckless and unusual sexual promiscuity, which are some of the symptoms of mania (Guha, 2014).

For example, when people are inside a nightclub, alcohol, with music, combined with flashing lights provide a social environment for mania and not for genuine joy. And people easily tend to confuse the state of genuine joy with euphoria and mania.

This is most noted with people who unconsciously seek to escape from negative emotions but do not generate authentic positive emotions. The confusion is due to the positive emotional intensity provoked by the craze, which gives rise to a spectacular short-term but not sustainable positive feeling, which is characteristic of mania (Guha, 2014).

This effect is similar to the trap of drugs, which cause a very intense positive feeling generated by the mania, effect of the chemical intake, that initially confuses the individual, making them want to experience the same high and intense positive sensation again. This causes dependency and compulsion, which subsequently form a dangerous and damaging vicious circle for the individual. Later in this book, the neural bases of addiction are explained in detail.

Clearly, mania is a false joy. Genuine joy is an emotional experience that must start from the real self, even if it is as a reaction to an extrinsic positive event. It needs to have a positive genuine resonance in the real self to be sustainable in time and frequency in order to generate genuine happiness (Lyubomirsky et al., 2005).

The joy that comes from pride also resembles mania more than joy. For example, humility is a virtue more favorable to the generation of genuine joy, and although it is less intense, it does not generate any illusions of grandeur, providing a more sustainable happiness in the long term (Peterson & Seligman, 2004).

The capacity to develop the practical distinction between genuine and frequent joy with moderate intensity from the false high intense state of mania or euphoria is an important characteristic of emotional intelligence, to look for genuine positive emotions and to avoid illusions that can lead to addictive behaviors or psychopathologies.

The Neural Bases of Joy

A meta-analysis composed of the statistical synthesis of 83 neuroscientific studies demonstrated that the emotion of happiness is selectively associated with the superior temporal gyrus (Vytal & Hamann, 2010).

Another meta-analysis composed of a statistical synthesis of 55 neuroscientific studies demonstrated that in 70 percent of these studies, the basal ganglia are related to the emotion of happiness, specifically the ventral striatum (which involves the nucleus accumbens [NAcc] and the putamen) (Phan et al., 2002). Researchers consider the NAcc and putamen as brain regions associated with the reward system that are closely linked to the feeling of pleasure in the brain.

Although researchers do not distinguish, in cognitive neuroscience, the concept of happiness/joy from the concept of pleasure, activations in the reward system usually are associated with the processing of rewards (Rolls, 1999), pleasurable activities (Koepp et al., 1998), responses after observing facial expressions of happiness (Phillips et al., 1998), pleasurable images (Lane et al., 1997), memories of happy events (Damasio et al., 2000), sexual pleasure (Rauch et al., 1999) and the joy of feeling progress toward a goal (Davidson & Irwin, 1999).

Regions belonging to the basal ganglia have many dopaminergic neurons and are therefore associated with the feelings of pleasure, joy, and happiness, and even with drug-addictive behaviors (Breiter et al., 1997).

Neural Mechanisms of Reward

The brain has what neuroscientists call a reward system. A reward is when someone reaches something that they want; therefore, that person tends to experience joy and pleasure.

The inner experience of joy when helping someone is quite different from the joy obtained from sexual pleasure or from having fun with friends. Despite this, several different types of "psychological rewards," such as donations to charities, eating a sweet, earning money or seeing a smile on a face activate similar brain regions, namely, the orbitofrontal cortex, the anterior cingulate cortex, the insula, the NAcc and the pallidum (Berridge et al., 2009).

There are two distinct mechanisms of rewards, namely, "liking" and "wanting" (Berridge et al., 2009).

Reward Mechanisms Associated With "Liking"

The reward mechanism associated with the idea of "liking" can be characterized by the joy generated through pleasure. Some researchers, for example, carried out experiments in rats using anandamide microinjections in the NAcc of rats, which made these animals feel much more delight in ingesting this substance and increased the tendency to consume twice the amount.

In humans, the ventral pallidum is activated as a result of several types of rewards, such as cocaine, sex, money or food, and it is another important region of the reward system in the brain, in addition to the NAcc (Berridge et al., 2009). For example, in humans, the reward anticipation, or the expectation to receive a reward, such as money, tends to activate the NAcc (Knutson et al., 2001).

Researchers have found some neurochemical systems, like those that increase the reaction of "liking" in rats, only in some very circumscribed subcortical regions of the brain, particularly within the NAcc. For example, opioid neurotransmitter systems, endocannabinoids and GABA-benzodiazepine are important systems for the generation of pleasure reactions, what researchers call "liking" (Berridge et al., 2009).

Reward Mechanisms Associated With "Wanting" (Impulse Toward Action)

As previously mentioned, when the person achieves something that they "want," they tend to "like" it. Wanting works as a motivational system; that is, it stimulates the impulse for action, which is the essence of the concept of motivation.

Research has shown that in the brain, the mechanisms of "liking" and "wanting" are dissociated (Berridge et al., 2009). For example, a group of researchers defined "wanting" as the incentive for action (Berridge et al., 2009), as if it was a behavioral impulse to desire something, different from a declarative and cognitive desire. It is more of a boost to action, or what we call impulse for action, rather than a psychological explanation of a desire (Berridge et al., 2009). As with addictions, for example, the addiction to drugs, it is even possible to "want" something without liking it. According to neuroscience, this is the essence of the concept of "wanting" (Berridge et al., 2009).

Addictions and compulsions are "irrational" ways of "wanting." In these cases, there is an impulsivity (i.e., the impulse to action) without self-control. Emotional intelligence allows human beings to develop strategies that help to control not only their own emotions but also their impulses of "wanting" or destructive compulsions.

Within the NAcc, there are two regions that are dissociated: one is responsible by "wanting" to consume the object of pleasure, while the other guarantees "liking" the object (Berridge et al., 2009).

When, irrationally, animals or humans activate the "wanting" reward system, they may develop problems of addictions, be it drugs, alcohol, or other substances. It is what researchers call an irrational and destructive "wanting." Initially, the neural process of addiction is characterized by an alignment between "liking" and "wanting," to the point that the reward system of the "wanting" overlaps with the reward system of "liking," and it is possible for the body (human or animal) to "want" something even without "liking" it. This is the case of drug addiction – a person "wants" the drug; however, he does not "like" to consume it anymore. This is the phenomenon of addiction (Robinson & Berridge, 1993).

In the gradual process of addiction, from the beginning, there is an alignment between the "liking" and "wanting" mechanisms. But, as the habit of compulsive addiction tends to increase, the "wanting" grows, while the "liking" progressively decreases.

That is why for human beings, it is very important to be careful with pleasures. They attract us (the initial "liking" reward mechanism) to "want" a set of choices and actions that, in the long run, neither do us good nor generate continuous happiness and well-being.

And we are not just talking about addictions.

False motivations and false values can generate illusory effects of pleasure, which does not necessarily promote evolution, complexity, or genuine psychological growth (Csikszentmihalyi, 2004, 1993).

Being emotionally intelligent also means having the capacity to prevent emotions and pleasures from guiding us indiscriminately. It is knowing how to make intelligent choices that drive our behaviors and emotions to opt for emotions and pleasures that enable health, happiness, and well-being to generate sustainable inner development to individuals and relationships.

Moral Emotions

There are two fundamental dimensions of moral emotions. The first is that moral emotions tend to motivate prosocial moral behavior. The second is that these emotions are also triggered in the face of moral violations. So, in one dimension, moral emotions stimulate positive moral behavior, while in another, they tend to create an aversive emotional system in response to moral violations (Haidt, 2003a).

Haidt (2003a) proposed the following classification for the families of moral emotions: those that enhance others (e.g., gratitude, admiration, and moral elevation); the ones that condemn others (e.g., contempt, anger, moral disgust, and moral indignation); the self-conscious emotions (e.g., shame, embarrassment, guilt, and pride); and the emotions felt in response to the suffering of others (e.g., compassion).

Emotions Toward the Suffering of Others

The main positive emotion in the face of the suffering of others is compassion, which is explained in detail next.

Compassion

Compassion is provoked by the perception of suffering or sadness in another person. It seems to grow from the mammal binding system, in which there are benefits as a mediator of altruism among human beings (Hoffman, 1982). You can feel compassion for complete strangers. However, this emotion is stronger and readily felt when it involves relatives or other people with whom one has a close relationship (Batson & Shaw, 1991).

Compassion makes people want to help, comfort, or alleviate the suffering of others (Batson & Shaw, 1991; Eisenberg et al., 1989; Hoffman, 1982).

It is very important to encourage children, adolescents, and adults to cultivate compassion, as well as the value of benevolence that leads to that emotion, because a great part of humanity's problems have their deepest cause in the lack of the practice of compassion or the dissemination of social and organizational values that ignore the suffering of others.

The cultivation of the moral emotion of compassion and the values that stimulate compassion, altruism and benevolence should occur in a wide range of personal contexts, interpersonal contexts and social issues.

Fortunately, in the United States, a movement has grown scientific, and there is a social stimulus to compassion, with scientific bases and applications in areas until recently considered unthinkable, such as compassion training for physicians (Leget & Olthuis, 2007; Patel et al., 2019), training in self-compassion (MacBeth & Gumley, 2012; Leary et al., 2007; Neff & Vonk, 2009; Barnard & Curry, 2011) as one of the mechanisms for reducing incidences of depression, training in economic and organizational decision-making with compassion (Singer & Ricard, 2015), compassion training for children and adolescents (Peterson, 2016) and training in compassion between couples (Gottman & Silver, 2015; Gottman, 2008), among others.

THE NEURAL BASES OF COMPASSION

The neural bases of compassion have been studied in parallel with the neural bases of empathy for pain, considered the evolutionary origin of compassion and altruism. In summary, the neural bases of empathy for pain involve activation of two main brain regions, the anterior insula and the anterior medial cingulate cortex.

The amazing thing is that neuroscientific research has already been able to demonstrate that the higher the activation of the insula, the greater the number of donations to organizations (Tusche et al., 2016). This is the same as saying that the activation of the anterior insula can predict prosocial behavior, with a scientific relation of cause and effect (Figure 1.7).

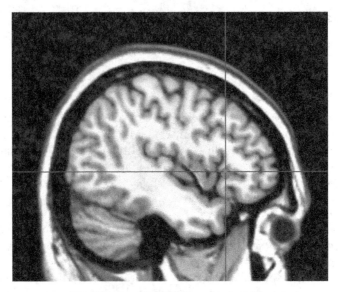

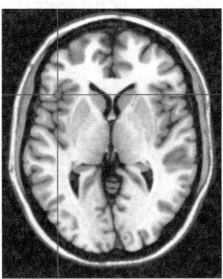

Figure 1.7 Representation of two images of the anterior insula associated with the emotion of compassion and a predictor of the prosocial behavior of donation, according to functional magnetic resonance imaging, at the intersection of the horizontal line with the vertical line.

Another important finding is the fact that neuroscientific research has also been able to identify the neural plasticity resulting from practical training in compassion, activated in specific regions (Klimecki et al., 2013).

The training in compassion consisted of a single day of training with the aim to help participants cultivate compassion for all human beings, with the control group composed of memorization tasks (Klimecki et al., 2013).

After the training in compassion, it was found that the regions of the area of the ventral/black substance, ventral striatum, NAcc, globus pallidus and putamen were activated (Klimecki et al., 2013). As mentioned previously, these brain regions are associated with the reward system, which means that compassion is an intrinsically rewarding moral emotion.

In addition to the previously mentioned regions, other regions were also activated during the training: the anterior subgenual cingulate cortex and the medial orbitofrontal cortex (mOFC).

As for the subgenual anterior cingulate cortex, previous studies had identified its relationship with empathic concern, i.e., the higher the activation of this region, the greater is the tendency toward empathic concern (Zahn et al., 2009).

And it is known that the latter is one of the most important characteristics of compassion, that is, to genuinely care for the well-being and happiness of others.

Regarding the mOFC, it makes perfect sense for this region to be activated. According to Professor Damasio, when patients had lesions in this area, they were associated with difficulties to self-generate moral motivation (Damasio, 2005).

Gratitude

Historically, gratitude has always been highly valued among Jewish, Muslim, Christian, Buddhist, and Hindu religions. In the history of science, until recently, gratitude has been a highly devalued object of scientific inquiry (Emmons & McCullough, 2004).

The main idea about gratitude is that it is an emotional response to a moral act of another person who has benefited us, and who generates the moral emotion of gratitude (McCullough et al., 2002).

In experimental psychology, gratitude is conceptualized as a moral emotion (Haidt, 2003a). In addition to being an emotional response, gratitude is a source of moral motivation, which impels us to perform acts of kindness and to develop prosocial behaviors in response to a benefit that someone has given us. In this respect, gratitude can be considered a source of moral motivation to do good (McCullough et al., 2002).

In science, the most eminent scientist to write about gratitude was Adam Smith, one of the founding fathers of modern economics. Paradoxically, the same man who wrote *The Wealth of Nations* (Smith, 2010), a work that formed the basis for the evolution of economic thinking and growth of the material wealth of contemporary economies, wrote a book called *The Theory of Moral Sentiments* (Smith, 1790/1976), in which he focuses in depth on gratitude and its importance in contemporary societies.

Smith proposed that human emotions are guides to judgments and moral decisions, including economic ones. In this context, Smith argued that gratitude is one of the most important moral emotions, and one of the main drivers of benevolence to someone who has previously benefited us somehow (Smith, 1790/1976). According to Smith, it is also one of the most important moral emotions to be cultivated in behavior economics, which, for him, should be based on goodwill. Smith claimed that a society based solely on economic utilitarianism, without considering the existence of moral emotions, would clearly, as a consequence, have many difficulties in maintaining socio-economic stability (Smith, 1790/1976). In fact, all these statements were confirmed during the 20th and 21st centuries.

Smith also presents three psychological factors that govern gratitude, i.e., a person experiences gratitude when they feel that the other person intended to benefit them, when the person was successful in benefiting them, or when the benefactor was able to tune in to the gratitude emotions of the beneficiary (Smith, 1790/1976).

For this philosopher, as soon as these three conditions are met, the experience of gratitude is fully lived by the beneficiary. Nevertheless, Smith also recognizes that, in practice, not all conditions can always be met, but they can generate experiences of gratitude (Smith, 1790/1976).

In addition to Adam Smith, Simmel (1950) and Gouldner (1960) conceptualized gratitude as a force that helps humanity to maintain the obligations of reciprocity in human relations. Simmel (1950) refers to gratitude as a "moral memory of humanity."

However, the most modern theory about gratitude originated from Michael McCullough and colleagues (McCullough et al., 2001). These researchers claim that gratitude is a moral emotion within an entire universe and taxonomy of moral emotions that exist in human beings (McCullough et al., 2001; Haidt, 2003a).

According to these researchers, the moral emotion of gratitude has three fundamental functions: (1) it is a moral barometer, (2) it is a source of moral motivation, and (3) it is a positive reinforcer to the moral behavior of benevolence.

Gratitude as a moral barometer means that gratitude is the perception that someone benefited us, and this indicates a sense of moral uplift of the benefactor, impelling us to feel a more positive connection.

The moral emotion of gratitude increases in relation to our benefactor when we have received a benefit that is important, when this benefit required considerable effort from our benefactor, when the effort expended by our benefactor was intentional and not accidental or when the benefit generated by our benefactor was realized without second thoughts and disinterestedly.

Gratitude as a source of moral motivation is another important aspect of this theory. The moral emotion of gratitude tends to impel people to act more benevolently toward others and motivate people to want to be better humans.

Researchers speculate that gratitude is one of the underlying mechanisms of reciprocal altruism – that is, doing good to a person impels them to do the same by the other or to a third person, as a form of reciprocity (McCullough et al., 2001).

In addition, gratitude tends to generate less destructive behavior stemming from the benefactor and others (McCullough et al., 2001). Another aspect that researchers refer to is that the thrill of gratitude is different from the thrill of being indebted to someone. While the thrill of gratitude makes people feel happier, the emotion of debt can lead people to feel in a position of discomfort in relation to the benefactor (McCullough et al., 2001).

Thus, the thrill of being in debt is a negative emotion, while gratitude is a positive emotion that can provoke the emotion of moral elevation and tends to increase happiness (Haidt, 2003a).

Gratitude as a positive reinforcer to the moral behavior of benevolence means that people expressing gratitude for a person who has been a benefactor in a particular moment tends to reinforce that person's behavior to acts of kindness in the future (Einsenberg et al., 1991).

Ingratitude to a benefactor tends to generate emotional reactions of anger and resentment, and does not encourage the benefactor to engage in future acts of kindness to the beneficiary.

Tesser and collaborators (1968) demonstrated the role of the moral barometer of gratitude and tested the hypothesis that three factors would be associated with the intensity of gratitude: the perceived intentionality of the benefactor, how costly the benefactor's action was to the benefactor, and the value of the benefit.

McCullough and collaborators (2002) found that people who cultivate gratitude tend to be happier, to more frequently feel positive emotions and to undertake more prosocial behaviors. These researchers also revealed that high levels of gratitude were associated with low levels of materialistic values and jealousy.

In a series of three experimental studies, Emmons and McCullough (2003) asked participants to list the blessings they received during several weeks (experimental condition of gratitude). Respondents demonstrated higher levels of happiness and optimism in life and reported more time to exercise gratitude and minor negative symptoms compared to a neutral condition (study 1).

In study 2, participants who filled out a gratitude journal felt more positive emotions than the neutral group. In addition, participants on the condition of gratitude tended to help and provide more emotional support to others compared with the neutral condition.

In study 3, the researchers demonstrated that in the condition of gratitude, participants showed higher levels of positive emotions, longer sleep time, better sleep quality, higher levels of optimism and higher social connection. It should also be noted that study 3 showed that the experimental group of gratitude obtained lower levels of negative emotions.

Thus, the studies concluded unequivocally that participants who cultivate the moral emotion of gratitude on a daily or weekly basis tend to significantly increase the levels of emotional well-being and happiness.

THE NEURAL BASES OF GRATITUDE

In one of the best neuroscientific studies of gratitude, the researchers identified, through the predictors of gratitude, a few important aspects, including the cost of action for the benefactor and the size of the benefit for the individual who received him as two of the greatest predictors of the moral emotion of gratitude (Yu et al., 2018).

Regarding the size of the benefit received, the brain region associated with this phenomenon was the right ventral striatum (Figure 1.8) (Yu et al., 2018).

It makes sense to activate this region, because the right ventral striatum is an important part of the brain reward system (Yu et al., 2018). According to this study, the greater the benefit received, the greater was the activation of this region (Yu et al., 2018).

Regarding the cost to the benefactor, the higher the cost to the benefactor, the higher is the activation of the MPFC and the precuneus (Yu et al., 2018).

As far as the emotion of gratitude is concerned, it activated the perigenual anterior cingulate cortex and was highly correlated with high levels of gratitude measured by a gratitude questionnaire as a personality trait (Yu et al., 2018).

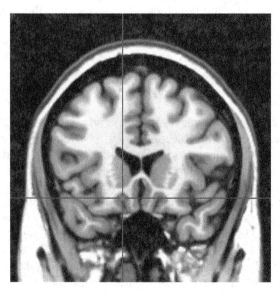

Figure 1.8 Representation of the right ventral striatum associated with the emotion of gratitude, at the intersection of the horizontal line with the vertical line.

The Neural Bases of Emotions   25

In addition to the activation of the anterior cingulate perigenual cortex associated with gratitude, activation of this region is also correlated with high levels of gratitude from a quantitative questionnaire, and it is even possible to verify that the higher the activation of this brain region, the greater is the tendency for individuals to donate money with the aim of repaying the benefit received (Figure 1.9) (Yu et al., 2018).

Professor Damasio's laboratory also identified that the brain regions associated with gratitude are the MPFC and the anterior cingulate cortex (Fox et al., 2015).

Moral Elevation

Moral elevation is triggered by the observation of acts of kindness, loyalty, self-sacrifice and extraordinary moral actions carried out by others.

Like gratitude, moral elevation triggers "affection for the person who provoked the emotion."

Moral elevation tends to create a widespread motivation that leads the individual to want to become a better person and to follow in the behavioral footsteps of the moral exemplar (Colby & Damon, 1992; Algoe & Haidt, 2009). People who experience moral elevation are more likely to help others, to donate money to charity and to behave prosocially (Algoe & Haidt, 2009; Haidt et al., 2002).

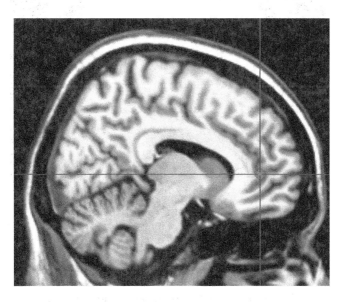

Figure 1.9 Representation of the perigenual anterior cingulate cortex associated with gratitude, at the intersection of the horizontal line with the vertical line.

Moral elevation opens people to new possibilities of thought and action, making them more receptive to the lessons of a moral specimen (Fredrickson, 2013). This process of mental and moral openness may explain why the narratives of lives of saints and religious leaders (e.g., Buddha, Jesus, among others) include so often reports of people who have changed to less materialistic lifestyles (Haidt, 2003b).

THE NEURAL BASES OF MORAL ELEVATION

One study found that as a response to a video that stimulates the emotion of moral elevation, the most synchronized brain region activations are the right anterior prefrontal cortex (Figure 1.10) (Englander et al., 2012).

It is understandable that this region is activated by the emotion of moral elevation because previous studies reveal an association between the right anterior prefrontal cortex and the stimuli of prosocial moral images (Moll et al., 2002).

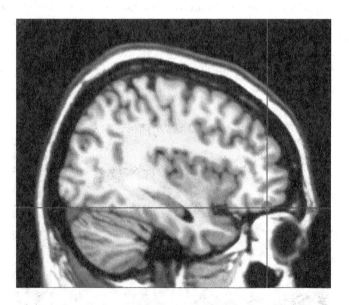

Figure 1.10 Representation of the right anterior prefrontal cortex associated with moral elevation, at the intersection of the horizontal line with the vertical line.

Admiration

Admiration is the emotion that people feel when they see extraordinary demonstrations of skill, talent or achievement (Algoe & Haidt, 2009). It is an emotional response to actions that demonstrate non-moral excellence, like an extraordinary talent, skill or competence, while moral elevation is the emotional response toward moral excellence (Algoe & Haidt, 2009; Haidt, 2003a).

Darwin described the admiration as a "surprise associated with some pleasure and sense of approval" (Darwin, 1965).

Admiration generates a desire for closeness to highly qualified and talented people, especially if these talents and abilities are achievable by themselves.

It also tends to generate strong motivations for energy, inspiration, achievement and psychological vitality (Algoe & Haidt, 2009; Haidt, 2003a).

THE NEURAL BASES OF ADMIRATION

Professor Damasio's laboratory also investigated the neural bases of admiration, an emotion that his team divided into two types: admiration for virtue and admiration for capacity (Immordino-Yang et al., 2009). The first happens when we admire a virtue of a certain person, and the second emerges as a reaction to a capacity of a person (Immordino-Yang et al., 2009).

Activated brain regions that are associated with admiration for virtue are the insula, the anterior cingulate cortex, the hypothalamus, the posterior cingulate cortex and the posterior inferior and posterior medial cortices (Immordino-Yang et al., 2009).

Regarding the admiration for capacity, the activated regions were the insula, the cingulate cortex, the dorsal posterior cingulate cortex and the posterior inferior and posteromedial cortices (Figure 1.11) (Immordino-Yang et al., 2009).

What the activations of the anterior insula and anterior cingulate cortex suggest is that admiration tends to activate a system of self-consciousness (anterior insula), with emotional characteristics, and a self-reflection (anterior cingulate cortex), not only at the level of virtues, in the case of admiration for virtue, but also at the level of talent, skills, capacities and competencies, in the case of admiration for capacity.

Moral Emotions That Condemn Others

The moral emotions that condemn others are anger, moral disgust and contempt.

Anger/Moral Indignation

Anger as a moral emotion that condemns others is an emotional response toward situations of injustice, lack of equity or serious moral violations from others, as well as unjustified insults or aggressions, especially when the latter

28 What Neuroscience Can Teach Us About Emotional Intelligence

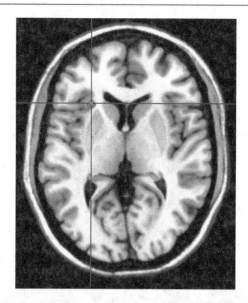

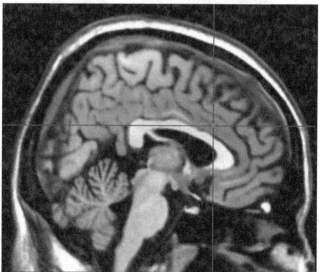

Figure 1.11 Representation of the right anterior insula and the anterior cingulate cortex, at the intersection of the horizontal line with the vertical line.

are unfair. Anger involves a motivation to attack, humiliate or turn against those who are perceived to act unfairly or immorally (Izard, 1977).

In some situations, anger is necessary, for example, in the face of aggression or excessive injustices, and it can cause essential changes at the individual or even the social level.

It is believed that the aggressiveness of the black communities in the United States in the struggle for their rights against white racism was necessary, otherwise, they would never have achieved civil rights in the United States.

Of course, unproductive and destructive outbursts of anger are not justified, not least because most of these explosions are not smart or productive. The problem is when anger becomes the dominant emotion in all conflict situations. From a certain point, anger proves to be highly destructive and, consequently, unproductive.

Therefore, one of the main roles of emotional intelligence is to manage the emotion of anger in an emotionally intelligent way. For example, if it is verified that the person or social group acts intentionally in a negative way and violates the most basic rules of morality, not listening to any peaceful solution, anger, as long as it is punctual, can be quite effective in forcing the other part to review its position and modify its behavior.

Unfortunately, in the history of humankind, many social changes would not have been possible without the use of anger and violence.

The dialogues between Albert Einstein and Mahatma Gandhi about the strategy of civil disobedience without violence, used by Gandhi in the struggle for independence from British forces, are famous. But this strategy would never have worked against the aggressiveness of Hitler in Europe during World War II. Only the aggressiveness of the American general Patton increased respect through fear from the German troops (Patton was the only American general feared by the Nazis), which allowed the victory of the Allies against the Germans in World War II (Einstein, 1954; Williamson, 2009).

Anger, when used strategically in extreme situations and with a purposeful intention, can be a useful emotion to correct a moral violation of a son or daughter during the educational process. It allows a shock of awareness to moral principles and appropriate behaviors in children and adolescents who do not have any intrinsic motivation to follow or to respect them.

The art is the ability to use it in an emotionally intelligent way, at the right time, in the right way, for the right reasons, with the right intensity (Aristotle, 2011). And this requires an accurate sense of self-control and educational strategy from the parents.

THE NEURAL BASES OF ANGER/MORAL INDIGNATION

See "The Neural Bases of Anger."

Moral Disgust

Moral disgust is a moral emotion that arises as a reaction to a moral violation committed by another person, such as a human trafficker or a pedophile.

While disgust is a basic emotion that arises as an aversive reaction to polluted objects to protect physical health, moral disgust, to preserve the intrinsic and individual system of morality, is an aversive reaction to moral violations to preserve the moral integrity (Izard, 1977), through an aversion to any serious moral violation perpetrated by other individuals.

If there were no system of aversion to immorality characterized by the emotion of moral disgust, we would probably be unable to preserve the integrity of our moral behaviors, and we would see ourselves "contaminated" or "infected" by immoral behaviors, due to the inability to distinguish right from wrong, from a moral standpoint (Haidt, 2003a).

Behavioral trends triggered by the emotion of moral disgust are the motivation to avoid, expel or, in any way, interrupt the contact with the entity that offended us or committed an immoral, corrupt, or negative deviant action, and the motivation to wash, purify or remove waste from any physical contact with the entity that offended us.

One study showed that people do not want to have contact with the clothes or other personal belongings of those who have a bad nature, such as a sweater worn by Adolf Hitler (Rozin et al., 1994).

THE NEURAL BASES OF MORAL DISGUST

According to investigators, moral disgust can also be called moral indignation, and it tends to activate the same brain regions associated with physical disgust, and specific brain regions associated with the moral dimension of disgust.

The main regions related to moral disgust are the Broca area or inferior prefrontal cortex – the anterior prefrontal cortex and the lateral orbitofrontal cortex (Figure 1.12) (Moll et al., 2005).

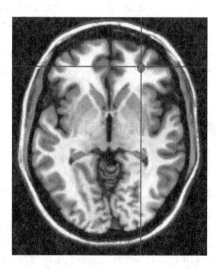

Figure 1.12 Representation of the left lateral orbitofrontal cortex associated with moral disgust, at the intersection of the horizontal line with the vertical line.

It is logical that this region is associated with moral disgust, because in another study, researchers identified it as being related to the withholding of donations to punish organizations that people considered harmful, such as the American National Rifle Association, among others (Moll et al., 2005).

Thus, this region was associated with so-called altruistic punishment. This study of the neural bases of moral disgust also suggests that this type of altruistic punishment may be linked to the emotion of moral disgust.

Contempt

The expression of contempt involves looking at someone with a sense of superiority in some domain, be it material, hierarchical, social, organizational, aesthetic, intellectual or moral (Izard, 1977). Contempt exists to mark and maintain classification distinctions and prestige. It is characterized by a cold indifference, an assertion that the other is not even worthy of strong feelings like anger.

Contempt does not motivate either attack or withdrawal; on the contrary, it seems to cause cognitive changes so that the object of contempt will be treated with less respect and consideration in future interactions (Fischer & Giner-Sorolla, 2016; Oatley & Johnson-Laird, 1996).

The emotion of contempt tends to weaken compassion for those who are the object of contempt (Haidt, 2003a).

THE NEURAL BASES OF CONTEMPT

In a study of the neural bases of contempt, researchers identified the amygdala as the main activated brain region (Sambataro et al., 2006).

The activation of the amygdala suggests that contempt is a more "refined" and "intellectualized" form of anger, being anger based. When you feel anger, you decide to find attributes that allow you to despise or diminish the attributes of another person.

And the activation of the amygdala allows that to happen (Figure 1.13). When investigators compared the condition of contempt with the condition of moral disgust, they identified that the amygdala is more strongly activated in the first case, which refers to the previously mentioned idea.

Self-Conscious Moral Emotions

Self-conscious moral emotions are guilt, shame, embarrassment and pride. They include the emotion of pride in the typology of positive emotions, which is described in detail in the Positive Emotions section, and not in conjunction with moral emotions.

Guilt

Guilt originates from the violation of moral rules (Freud, 1930/1961; Lazarus, 1991), especially if they cause harm or suffering to others (Hoffman, 1982).

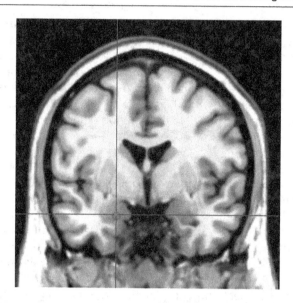

Figure 1.13 Representation of the right amygdala associated with contempt, at the intersection of the horizontal line with the vertical line.

Typically, this emotion can be characterized as a feeling of anger toward oneself (Westerink, 2009) or sadness toward oneself.

If it is prolonged in time, it may lead to a negative overall assessment of the self and negatively affect the self-esteem, leading to depression (Green et al., 2013, 2012).

It can also be triggered if the harmful action creates a threat to the established communion with the victim. Guilt can still be solitary, such as breaking a diet, masturbating or because of an evil done to another person without anyone knowing about it. This self-conscious moral emotion can motivate people to victimize themselves or compensate for the transgression. In some situations, the guilt can also encourage people to treat their partners well or to ask for excuses and confess their mistakes, as a way of restoring or improving the quality of their relationships (Haidt, 2003a).

Guilt allows us to evaluate the quality of an action in the face of a moral reference and tends to prevent the criminal or destructive intentions of oneself toward others. Therefore, guilt is a fundamental moral emotion to the moral evolution of human beings, as it avoids an escalation of destructive thoughts and actions toward others. Studies on sociopathic patients show that the suppression of the moral emotion of guilt is one of the characteristics that leads criminals to continue to commit crimes (Thorne, 1959).

In addition, it helps to enforce the moral norms of concern, respect and positive care toward others, even in the absence of immediate personal return.

In addition, it motivates people to pay attention to and express positive feelings toward others (Baumeister et al., 1994).

By allowing a person to convince another person to do what is considered correct, this emotion can also function as a form of interpersonal influence and persuasion. If person A wants person B to do something, A can induce guilt in B, transmitting how A suffers from B's inability to act in the desired way.

Person B finds guilt aversive and, to escape it, complies with the guidelines of A. It is clearly a type of influence very common in intimate relationships because A is not offering reciprocal benefits in terms of appropriate behavior, and, to convince the other to change his behavior, uses guilt to create an affective state that will motivate B to do what A wants to do (Baumeister et al., 1994).

Thus, guilt, if it is strategically expressed in specific situations of transgression on the part of the individual, functions as an emotion that enables the regulation of human morality, preventing human beings from committing moral violations uninterruptedly. This is the emotionally intelligent way to manage emotion moral guilt: allow an in-depth analysis of the moral behavior of humans.

The problem is when guilt becomes chronic. Several studies show that when guilt becomes chronic it tends to lead to depression. When guilt generalizes to the global sense of self, it tends to be extremely self-destructive (Green et al., 2013, 2012).

Generalized blame for all aspects of the self is not an emotionally intelligent way to manage the emotion of guilt. Emotional intelligence allows regulation, both avoiding of the absence of guilt (which can lead to sociopathy) and its excess generalized to the global sense of self (which can lead to depression). Again, as Aristotle (2011) wisely suggested, the middle way is the way of the wise man. The absence of guilt is as harmful to the human being as its excess.

Shame

Shame is triggered by the assessment that there is something wrong or defective with one's own behavior, usually due to the lack of morality measures, aesthetics or competence, or in response to a violation of cultural norms or social conventions (Tangney et al., 1996). The function of shame is to be a regulator of how one should act, but this expands to regulate how one should behave as a human being (Haidt, 2003a).

It is said that shame is a painful emotion resulting from actions that reveal that a person is imperfect or defective (Lewis, 1971). As a rule, shame is caused by a perceived violation of a social norm (Tangney et al., 1996). Shame leads people to reduce their social presence by creating a motivation to hide, retreat or disappear, making movement and speech more difficult and less likely.

This emotion inhibits assertive behavior and signals that the individual recognizes the occurrence of a violation, thereby reducing the likelihood of attack or further punishment of other people with social power (Haidt, 2003a).

When compared to embarrassment, shame involves a more painful will to withdraw, which may even motivate suicide (Durkheim, 1987/1951). It should be noted that excessive shame may be due to excessive pride. Pride hurt in the face of too much concern about negative evaluation of others can be a strong generator of shame. Emotional intelligence means to reduce the excess of concern for other people's assessments of you but maintain an intrinsic and non-social moral reference on one's own behavior.

Embarrassment

It is known that shame is an emotion of greater intensity than embarrassment (Tangney et al., 1996); however, there are other important differences. Some researchers suggest that shame results from moral transgressions or social norms compared to the transgressions that trigger embarrassment (Lewis, 1992; Ortony et al., 1988; Tangney et al., 1996).

For them, embarrassment stems from trivial social transgressions of everyday life (Lewis, 1992; Ortony et al., 1988), for example, a slip and fall into a puddle of water in front of a bus stop full of people.

When someone feels embarrassment, they tend to blush, smile disconcertingly or feel "stupid" in a given situation, and unlike guilt, it is less likely to feel repentance or depression (Buss, 1980).

It is concluded, therefore, that while shame relates to more serious moral violations, embarrassment is only associated with violations of trivial social problems, which do not entail any kind of moral implication (Tangney et al., 1996; Buss, 1980; Lewis, 1992). Another difference is that shame is an emotion that can be felt in private, whereas embarrassment tends to be felt more frequently in a public situation (Tangney et al., 1996; Edelmann, 1981).

Embarrassment is also related to hierarchical interactions: a person who feels more easily embarrassed when they are close to someone with a superior social status is less likely to experience embarrassment among people of inferior status or with whom they have intimate relationships (Miller, 1996).

There is a tendency for humans to behave socially and present the appropriate "face" in certain social situations, particularly when they are in the presence of members of higher status or a prestigious social group (Goffman, 2017).

Although embarrassment tends to be a less intense emotion compared to shame, it is important to note that in embarrassment, the mechanism of the underlying pride is similar to the emotion of shame. They are both the emotional reactions of how others evaluate us socially.

Thus, the higher the intrinsic and non-social moral reference, the lower is the tendency to feel shame or embarrassment.

THE NEURAL BASES OF GUILT, SHAME AND EMBARRASSMENT

One of the best neuroscientific studies on the neural bases of guilt compared guilt with the emotions of shame and sadness in order to differentiate the specific brain mechanisms of guilt (Wagner et al., 2011).

These researchers identified a selective role of guilt in the activation of the right orbitofrontal cortex, in addition to a high correlation of this brain region with a quantitative measure of guilt as an emotional trait (Figure 1.14) (Wagner et al., 2011).

The fact that the emotion of guilt is related to the orbitofrontal cortex helps to support Professor Antonio Damasio's already classic idea that patients with brain injuries in this region tend to feel no guilt or notion of inappropriate social behaviors (Damasio, 2005).

In a recent meta-analysis (a statistical synthesis of 21 neuroscientific studies using functional and structural magnetic resonance imaging), the researchers identified a stronger correlation between guilt and the vACC, the posterior and precuneus regions (Bastin et al., 2016).

This same study recognized that shame is associated with the dorsolateral prefrontal cortex, to the posterior cingulate cortex and to the sensory-motor cortex (Bastin et al., 2016). It is interesting that there is a relationship between the dorsolateral prefrontal cortex and shame, because previous studies identified an association of this region with compliance with social norms (Spitzer

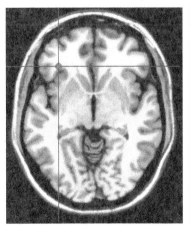
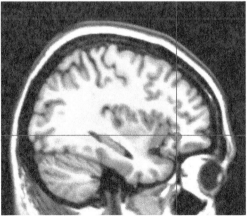

Figure 1.14 Representation of the right orbitofrontal cortex associated with an emotional trait of guilt, at the intersection of the horizontal line with the vertical line.

et al., 2007); therefore, shame, which is an emotion generated by the perception of violation of social norms, is also correlated with this region.

In the same meta-analysis, the researchers also identified a correlation between embarrassment and the ventrolateral prefrontal cortex (VLPFC) and the amygdala, suggesting that a person may fear the disapproval of others if they are in a social situation prone to the emotion of embarrassment (Bastin et al., 2016).

The activation of amygdala suggests that some underlying mechanisms of embarrassment may be associated with the fear of the social judgment of others.

Other Moral Emotions

Schadenfreude

The emotion of Schadenfreude is conceptualized as the pleasure you feel before the misfortune of others and contains an important moral component to the extent that it is stronger when the injured person is considered unworthy of their previous status (Portmann, 2000). Apparently, Schadenfreude does not involve a prosocial behavior tendency (Haidt, 2003a).

THE NEURAL BASES OF SCHADENFREUDE

One of the most decisive scientific pieces of evidence of the Schadenfreude phenomenon is when something bad happens to someone, the person feels pleasure. This is confirmed by a published study in *Science* in 2009, in which it was demonstrated that when a person feels Schadenfreude, this activates the ventral striatum, one of the brain regions most studied from the reward system and associated with the feeling of pleasure (Takahashi et al., 2009).

Envy

Despite being an extremely negative emotion from an intrapersonal point of view, and potentially very destructive to human relationships, envy is considered one of the most universal and present emotions in human beings (Russell, 1930). The tendency to feel envy is widespread and seems to be evident in all cultures (Foster, 1972; Schoeck, 1969).

Envy tends to be an unpleasant emotion, often painful, characterized by feelings of inferiority, hostility and resentment produced by being conscious that a person or a group of people have something that is very important for yourself, for example, an object, a social position, wealth, a personal attribute, a talent, a competence or a quality (Parrott, 1991; Parrott & Smith, 1993).

This mixture of feelings of inferiority, hostility and resentment generally persists in most definitions, although envy still has more complex contours,

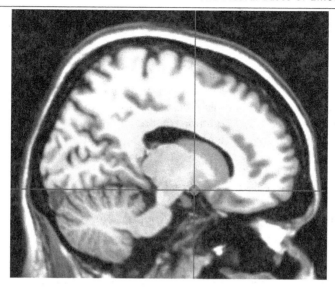

Figure 1.15 Representation of the ventral striatum associated with Schadenfreude, at the intersection of the horizontal line with the vertical line.

such as anger or sadness (Parrott, 1991). This emotion is also considered as a negative affective response in relation to the superior characteristics of other people (Ortony et al., 1988).

Why are we jealous? Why does a person's advantage make other people feel this emotion, which is painful from an intrapersonal point of view and very negative from a social point of view? Perhaps the most enlightening answer to this is that the advantages that other people often enjoy have strong consequences for the individual's self (Festinger, 1954; Mussweiler, 2003; Smith et al., 1989).

The relative position generally contributes strongly to deciding who wins what has value for the person. Social comparison also helps to form the basis for inferences about the self (Festinger, 1954). They also contribute to evaluation skills, for example, superior relative performance indicates success and high capacity, and lower performance indicates failure and low capacity (Kelley & Michela, 1980). This is particularly evident in the professional world and inside organizations and companies.

Due to the clear and obvious phenomenon of social comparison of performance within organizations, where, often, one competes explicitly or implicitly to achieve the best possible position, envy is a frequent emotion, together with its enormous social and moral negative consequences. Obviously because it is a highly undesirable emotion from a social point of view, the manifestation of envy is expressed in a fully veiled way. However, there are a number of identifiable symptoms for those who feel envy in organizations, such as intense fear or anger when a person is confronted with a high-performance achiever, such as a close colleague. This results from a perception of ego-threat and being harmed by high-performance achievement by a colleague. Interpersonal behaviors also come into play, such as in a conspiracy to attempt to diminish the reputation of the target of envy within an organization, which is a common phenomenon.

Humans tend to envy those in similar domains (Parrott, 1991). In addition, people tend to be jealous of others in domains that are very relevant to the self (Salovey & Rodin, 1991).

Envy differs from jealousy in that jealousy involves a third person, and one is afraid of losing that person to another (Foster, 1972; Parrott & Smith, 1993).

The confusion between these two emotions stems from the fact that in social life and media, there is a tendency to mix these two concepts. Dutch researchers were able to experimentally differentiate two types of envy: the benign, which leads to a motivation for change to improve your own position, and the evil, which drives the pull-down of the object of envy, to damage their position (Van de Ven et al., 2009).

It is crucial to teach humans to manage the emotion of envy in an emotionally intelligent way. The first way to do this is to train people in emotional intelligence, whether in the school context, in universities, in couples therapy or in family, or through training in organizations. First, what is important is to help people become emotionally aware of their envy, even if it is a socially negative and unpleasant emotion. When a person is feeling envy, it is necessary to admit and accept it, to allow it to be intelligently managed.

After the process of self-awareness, it is necessary to help people regulate the emotion of envy, and the degree to which they feel it. For example, if someone is feeling evil envy, at first, it is necessary to help them to transform this emotion into benign envy, and then to admiration.

This is the first phase of an emotionally intelligent regulation of envy. After this transformation, another intelligent way to manage this emotion is to transform it into a genuine admiration for the positive characteristics of the other person.

The ability to appreciate, affirmatively, the positive characteristics of others is a good sign of healthy emotional self-regulation of envy. Another very important aspect is to prevent thinking that the quality appreciated in someone else threatens our personal value and does not pose a threat to the self. It means that the quality of others does not need to be a threat to our own sense of self-value.

Our own self also needs to be appreciated in a very clear and affirmative way, not as an emotional reaction to other qualities but as a self-positive reference. Therefore, the positive psychology approach to virtues is an intelligent way to manage the emotion of envy. When we learn to appreciate the qualities and virtues inherent in ourselves, the qualities and virtues of others do not constitute a threat to ourselves (Peterson & Seligman, 2004; Niemiec, 2017). Intrinsic security from an appreciative, positive, and realistic sense of our real self makes us learn to appreciate the qualities of others without, therefore, feeling intimately threatened; it prevents the future possible emergence of envy. All these issues need to be verbalized naturally during training in emotional intelligence in organizational and personal contexts.

It is important to teach people how to manage intelligently the emotion of envy, both for the well-being of the one who feels the emotion of envy as well as to prevent it from generating destructive behaviors toward others.

THE NEURAL BASES OF ENVY

One of the most incredible findings is the scientific demonstration obtained by neuroscience that, effectively, when someone feels envy, they tend to activate the regions associated with social pain and conflicts between self-image and information (Takahashi et al., 2009).

Thus, envy provides feelings of inferiority, such as emotional conflicts between self-concept and the fact that someone is seen as superior in some attribute, i.e., social, and emotional pain (Takahashi et al., 2009).

The scientific evidence for this is the activation of the dorsal anterior cingulate cortex associated with the emotion of envy, when individuals were compared with someone who possesses high qualities and attributes, and simultaneously has high personal relevance for participants (Takahashi et al., 2009).

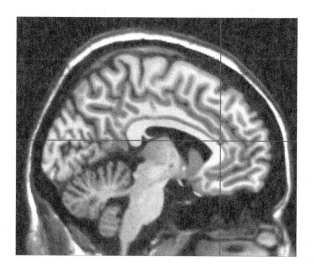

Figure 1.16 Representation of the dorsal anterior cingulate cortex associated with envy, at the intersection of the horizontal line with the vertical line.

Positive Emotions

The main positive emotions studied scientifically are joy, gratitude, serenity, interest, hope, pride, fun, inspiration, admiration and love (Fredrickson, 2013).

This list of positive emotions is organized in order of frequency in which they occur in humans, except for love, which will be the last emotion to present, despite being a positive emotion reported as very frequent (Fredrickson, 2013).

After explaining each of these ten emotions, the effects of positive emotions on negative emotions are presented as well as a theory that explains the mechanisms of positive emotions and their difference from negative emotions.

Joy

Joy is an emotion that emerges most frequently as an emotional response to a positive event (Watkins et al., 2018). But it is also possible to generate joy intentionally, regardless of external circumstances (Mathewes, 2015). Scientific studies suggest that those who have higher levels of gratitude in daily life tend to generate joy more frequently (Watkins et al., 2018). Many scientific approaches have emphasized the experience of joy after the occurrence of something positive that the individual expected or had hoped would happen a long time ago (Lewis, 1955; Vaillant, 2008).

Likewise, joy happens when we intentionally receive something good, and we feel that it is more than what we deserve (Volf, 2015). Also, when we are presented with something more positive than what we expected, we tend to feel the emotion of joy (Frijda, 2007).

This emotion can also occur because of an intrinsic motivation that meets the human needs of autonomy, competence, and affiliation, transformed into action, which is present in children and in intrinsically motivated adults or in those with intrinsic values stronger than extrinsic values (Deci & Ryan, 1985; Ryan & Deci, 2000).

Children tend to generate joy with ease because they are in a phase of rapid development of their cognitive, motor and emotional skills (satisfaction with the need for competence) and because they play and fantasize during the experience of autonomy (satisfaction of the need for autonomy). This occurs in a healthier environment when parents stimulate a secure and warm relationship (satisfaction of the need for affiliation or relatedness).

These motivational components, combined with the act of playing, stimulate the greatest frequency of emotion of joy, compared to adults. That is why the most innovative companies in Silicon Valley in Palo Alto, California, promote various games as practical activities within the organizations, to increase the frequency of positive emotions in its employees, and, consequently, increase the levels of innovation. For more details on the effects of positive emotions, see Chapter 2.

One of the most common effects of joy is the natural behavioral tendency to play, whether on an individual level or in relationship with others. This trend is present in both animals and humans (Fredrickson, 2013) and generates a sense of freedom resulting from increased psychological vitality.

An extraordinary scientific study demonstrated that the joy and enthusiasm of teachers while teaching their students tend to generate in them an increase in intrinsic motivation and psychological vitality to learn (Patrick et al., 2000). Thus, joy is an emotion that provides the enrichment of the quality of friendships, loving relationships and even interactions between coworkers, and also affects people at cognitive and mental levels, as presented in Chapter 2.

Until recently, in the world of work, there was a tendency to believe that professional people should be more serious, influenced by German and French labor cultures. In the 1990s, a former classmate of mine, who studied in Paris, was warned by teachers because of her expression of joy, which was the enemy of seriousness and professionalism!

Fortunately, in Europe, this preconceived idea is changing, although more slowly compared to in the United States. In this regard, see the commentary by Albert Einstein in his autobiography (Einstein, 1954).

Joy is also generated when a person feels progress toward their personal and professional goals, called the Progress Principle (Amabile & Kramer, 2011).

Joy is considered, with scientific bases, an important resource for creative work, organizational innovation, well-being and high-quality relationships (Fredrickson, 2013). Finally, joy is an emotion that results from the perception that the individual's life purpose is being achieved (Volf, 2015).

The Neural Bases of Joy

A meta-analysis of 15 neuroscientific studies found that joy is correlated with the anterior cingulate cortex, the prefrontal cortex, and the insula, although it cannot be said that joy has specific regions in the brain, because these areas are also activated by the emotions of sadness and anger (Suardi et al., 2016).

Gratitude

For more details on the emotion of gratitude, see section "Moral Emotions that Elevate Others – Gratitude."

Serenity

Also called contentment, serenity emerges when interpreting circumstances lived as totally satisfactory. People feel serenity when they are comfortable in each situation. Serenity stimulates assessment of the present circumstances and integration of them into new priorities and values.

The enduring resources generated by the emotion of serenity include a positive sense of self and a sense of clear priorities in life (Fredrickson, 2013).

It is also possible to feel serenity in the face of adverse situations, when the individual reaches a high level of inner security and psychological and emotional maturity, and that is a characteristic of wisdom.

Interest

The emotion of interest happens when one realizes that something new inward or outward is worthy of being explored and is important for the self. People develop this emotion when they come across something mysterious or challenging, and that looking to get to know it better can bring an important intrinsic or extrinsic gain (Fredrickson, 2013).

Interest generates the behavioral tendency to explore, investigate, learn and immerse in the self with the desire to deepen and expand it (Izard, 1977).

Science was born from the emotion of interest in understanding a particular phenomenon in depth, in whatever scientific domain.

Einstein's quotes about his interest and curiosity in understanding the laws of the universe are a fantastic expression of Einstein's frequent emotion of interest (Einstein, 1954).

According to Einstein, interest and curiosity were the true sources of intrinsic motivation, to study and understand the laws of nature (Fredrickson, 2013).

And it was the greatest joy for him when he could properly understand these laws (Einstein, 1954). Interest is a positive emotion that gives rise to a lasting resource for motivation in everything we do. Interest generates intrinsic motivation, psychological vitality and joy (Fredrickson, 2013).

Hope

Unlike the main positive emotions that occur in favorable circumstances, hope emerges from situations where people fear the worst will happen (Lazarus, 1991). The positive emotion of hope arises when a person realizes that, in the face of a negative situation, there is the possibility of a positive outcome and that in the future everything will improve.

Hope creates the predisposition of individuals to focus on their capabilities and to focus their efforts on trying to reverse an unfavorable situation. Hope builds the enduring resources of optimism and resilience to adversity.

People who are capable of generating the emotion of hope in a systematic and constant way tend to be more resilient and have a greater ability to reverse an unfavorable situation.

Without the emotion of hope, which is an emotion designed for the future, human beings would be unable to overcome problems and adversities and would end up crushed emotionally due to the problems and difficulties of life.

The Neural Bases of Hope

A neuroscientific study identified the role of the orbitofrontal cortex associated with hope and its protective role in anxiety and the ability to generate spontaneous brain behaviors to "distract" the mind from anxieties and to be constructive in the face of adversity (Wang et al., 2017).

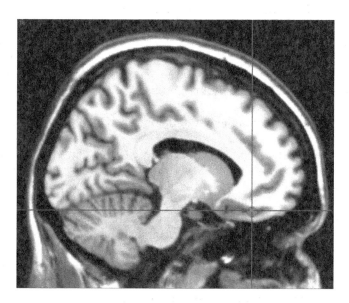

Figure 1.17 Representation of the medial orbitofrontal cortex associated with hope, at the intersection of the horizontal line with the vertical line.

Pride

Aristotle (2011) considered pride a virtue. For him, a truly proud human being should be good. And greatness in all virtues seems to be characteristic of a proud man (Aristotle, 2011). Aristotle, in his book *Rhetoric* (Aristotle, 1926), also distinguished the pride of arrogance and considered that this is not a virtue. About arrogance, he said, "As for pleasure in arrogance, its cause is this: men think that at the mistreating others makes their own superiority greater" (Aristotle, 1926). "That is why the young and the wealthy are arrogant. For they think that in being arrogant they are superior" (Lord, 2013).

Not only Aristotle but also some Greek thinkers used to have long discussions on the negative implications of arrogant behavior (Fischer, 1992). In addition, some Greek philosophers considered that pride and arrogance are two completely distinct concepts (Fischer, 1992).

Aristotle (2011) stated that a "virtue must have the quality of pointing to the intermediary [. . .] who cares about passions (now called emotions) and actions and in these there is excess, lack and middle term." He defended the idea that "feeling them (emotions) at the right times, with reference to the right objects, the right people, with the right reason, and on the right track, is what is simultaneously intermediate and the best, and this is a characteristic of virtue" (Aristotle, 2011).

He argued that this same idea can be applied not only to emotions but also to actions. In this sense, and following Aristotle's intermediate idea, we can define pride as an intermediate self-worth. An excess of self-worth would be arrogance, and its absence would be self-humiliation.

In addition to Greek thought, in most religions, such as Judaism, Islam and Hinduism, this distinction does not exist. Pride is seen as a negative attribute of the soul, is considered arrogance and, therefore, should be avoided.

In Christian tradition, pride is considered one of the seven deadly sins (Psalm 10:4; Bible, 2003). St Thomas Aquino wrote that self-love is the cause of every sin (Clark, 2000).

Different from all other religious traditions, the Dalai Lama stated:

'For example, when one has a distorted view of himself, as because of pride arrogance, due to these states of mind, there is a sense of exaggerated their personal qualities and abilities. Your vision of your abilities goes far beyond your real abilities. On the other hand, when you have low self-esteem, its real qualities and capabilities are underestimated. Belittled get down. This leads to a complete loss of faith in yourself. So, the excess, both in terms of exaggeration, and devaluation are both equally destructive. It is by addressing these obstacles and constantly examining their personal character, qualities, and abilities, which one can learn to have greater self-awareness. This is the way to become more aware.'

(Lama & Cutler, 2003)

It seems that like Aristotle's concept of excess, defect and intermediate, the intermediate pride is also considered positive from a Buddhist perspective (Lama & Cutler, 2003).

Scientists have also distinguished between achievement-oriented pride ("authentic" pride) and pride based on the distortion of the ego (Tracy & Robins, 2004). For these researchers, achievement-oriented pride ("authentic") is positive for both individuals' self-esteem and society, while pride based on ego distortion is negative for both, may generate mental illnesses such as narcissism, as well as may impair the quality of relationships with others. Do you know someone who likes to relate to someone with a narcissistic personality?

It is important to interpret the term "authentic" of these researchers as resulting from authentic achievements, and not derived from fantasies of the ego, such as occurs in pride-based distortions of the ego (Tracy & Robins, 2004).

According to the model of self-conscious emotions, people feel proud when their attention is focused on the self, activating representations about themselves and evaluating events congruent with their identity objectives and relevance to the self, and make internal attributions to the cause of events (Tracy & Robins, 2004).

Pride can also play a motivational role in perseverance in pursuing goals (Williams & DeSteno, 2008), and tends, as a result, to cause an increased level of optimism, further clarification of life purposes, a greater focus on personal goals, an active posture in the approach to life and an increase in self-regulation (Grant & Higgins, 2003).

This emotion arises when people take credit due to some good result valued or when they achieve an important goal (Tracy & Robins, 2007). It stimulates the will to fantasize about greater achievements than the ones already achieved, like big dreams (e.g., "authentic" pride; Williams & DeSteno, 2008).

Thanks to its benefits, it is important that educational processes of parents and educators encourage "authentic" pride and regulation of their excesses in children and adolescents. There are many parents who encourage excessive pride in their children, which makes their children distort the reality of their self through pride-based distortions of the ego.

This leads them to difficulties in dealing with their own weaknesses, and they tend to fall into the emotional processes of anger and sadness, which can generate depression in the long term. There is a need to increase awareness of children's development of authentic pride, and not ego-based pride.

There is an increasing number of incidences of depression in adolescents and young adults with low levels of resilience and little tolerance for confronting their own weaknesses. In part, this results from pride-based distortions of the ego, due to inadequate education by parents and teachers that tended to unconsciously cultivate ego-based pride in them, never encouraging them to observe their own weaknesses realistically and naturally.

During the self-building process, which occurs throughout life, the contrast between the ideal self and the real self is always an important educational component, because it is by contrast (what researchers call cognitive dissonance) that one increases self-awareness (Festinger, 1957; Harmon-Jones & Mills, 2019).

Escaping the confrontation arising from the contrast between the real self and the ideal self tends to prevent children, adolescents and even adults from increasing emotional self-awareness.

Excessive pride makes humans fragile to dissonant information between what people think they are and what they do, and everyone should view this contrast as something natural in the process of human development, and as a healthy process of self-awareness to achieve emotional maturity and positive change (Festinger, 1957; Harmon-Jones & Mills, 2019).

Encouraging excessive pride is making fragile humans in the face of information about the real self, preventing their progress in emotional intelligence. The stimulus to self-regulation of pride, whether in absence or in excess, is something fundamental in the process of emotional intelligence education, but it is a rather neglected aspect in the education of parents and schools, and even adults.

What can be concluded, either from the information from the Greek philosophers and of Buddhism or by the data coming from scientists, is that pride, provided with balance (neither absence nor excess), can be a positive emotion that generates motivation for achievement.

The challenge is to regulate the emotion of pride to the point of preventing the distortion of the vision of the real self, thereby preventing compromised mental and emotional health and the quality of relationships with others.

The Neural Bases of Pride

A group of academics led by a researcher at the Stanford Center for Compassion and Altruism Research and Education identified the emotion of pride as correlated with the posterior medial cortex, a region associated with self-processing (Simon-Thomas et al., 2012). Another discovery is that the greater the emotional experience of pride, the greater was the occurrence of deactivation of the dorsolateral prefrontal cortex (Figure 1.18) (Simon-Thomas et al., 2012). This is the most dorsal and lateral region of the prefrontal cortex.

Although several previous studies have identified this region as important for emotional self-control (Ochsner et al., 2002), it is interesting to note that these new data suggest that pride can lead to a decrease in emotional self-control. On the other hand, the experience of compassion tends to have the reverse effect, i.e., to activate this region (Simon-Thomas et al., 2012), suggesting that compassion can lead to an increase in emotional self-control.

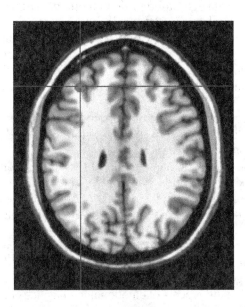

Figure 1.18 Representation of the dorsolateral prefrontal cortex, at the intersection of the horizontal line with the vertical line.

Fun

Fun occurs when people assess their current circumstances as involving some type of not serious social incongruity. It may arise when, for example, someone makes a harmless language error. Fun creates the desire to share a laugh and to find creative ways to perpetuate joviality.

As people follow these impulses, they build and solidify bonds (Gervais & Wilson, 2005).

The 20th and 21st centuries were characterized by a substantial increase in entertainment in the social life of Western societies and, though more recently, eastern societies marked by a Westernization of their cultures, as is the case of Asian countries. This feature of strong investment in entertainment and the cultivation of fun did not occur until after the Middle Ages.

On the contrary, during this time the most frequent emotions in Western societies, particularly in Europe, were fear and guilt. The phenomenon of entertainment and stimulating fun is relatively recent from the point of view of human history.

Fun tends to bring an emotional perception of greater lightness in how one lives day by day and a substantial increase in the frequency of positive emotions, which, from a psychological and social point of view, is very beneficial. When properly applied to the world of work, fun can generate better relationships between colleagues and stimulate the production of creative ideas, as well as generate faster problem-solving, higher levels of cooperation in negotiations and increased persuasive capacity in commercial teams. The only important aspects to regulate regarding the excitement of fun are to insert fun only in organizations that already have a serious culture of work and results and insert fun in relationships if this emotion does not stimulate the superficiality of values underlying social relations. The most innovative companies, particularly in Silicon Valley, tend to promote fun in work to stimulate organizational innovation, well-being, joy and interest.

Inspiration

Inspiration arises when human excellence is witnessed in some way. For example, people feel inspired as soon as they see someone performing a good deed or performing well, whether technical or behavioral and in whatever area it is. Inspiration triggers the will to overcome one's own personal and professional limits. The enduring resource that the emotion of inspiration stimulates is motivation for personal growth (Algoe & Haidt, 2009; Thrash & Elliot, 2004).

Admiration

For more details on the thrill of admiration, see section "Moral Emotions That Elevate Others – Admiration."

Love

Love arises when any of the positive emotions are felt in the context of a high-quality connection in relationship with others (Fredrickson, 2013). This emotion broadens the repertoires of thought-action, creating momentary perceptions of social connection and expansion of the self and its union with others (Waugh & Fredrickson, 2006).

Likewise, love builds a wide range of social resources that tie in with the community. The frequency of the emotion of love substantially increases the levels of emotional well-being and happiness, and the high-quality connection in human relationships (Fredrickson, 2013; Waugh & Fredrickson, 2006).

References

Adolphs, R., Tranel, D., Damasio, H., & Damasio, A. (1994). Impaired recognition of emotion in facial expressions following bilateral damage to the human amygdala. *Nature, 372*(6507), 669–672. https://doi.org/10.1093/neucas/3.4.267-a

Algoe, S. B., & Haidt, J. (2009). Witnessing excellence in action: The "other-praising" emotions of elevation, gratitude, and admiration. *The Journal of Positive Psychology, 4*(2), 105–127. https://doi.org/10.1080/17439760802650519

Amabile, T., & Kramer, S. (2011). *The progress principle: Using small wins to ignite joy, engagement, and creativity at work.* Harvard Business Review Press.

Aristotle (1926). In J. H. Freese (Trad.), *Aristotle in 23 Volumes* (vol. 22). Harvard University Press.

Aristotle (2011). *Nicomachean ethics.* The University of Chicago Press.

Barnard, L. K., & Curry, J. F. (2011). Self-compassion: Conceptualizations, correlates, & interventions. *Review of General Psychology, 15*(4), 289–303. https://doi.org/10.1037/a0025754

Bastin, C., Harrison, B. J., Davey, C. G., Moll, J., & Whittle, S. (2016). Feelings of shame, embarrassment and guilt and their neural correlates: A systematic review. *Neuroscience & Biobehavioral Reviews, 71*, 455–471. https://doi.org/10.1016/j.neubiorev.2016.09.019

Batson, C. D., & Shaw, L. L. (1991). Evidence for altruism: Toward a pluralism of prosocial motives. *Psychological Inquiry, 2*(2), 107–122. https://doi.org/10.1207/s15327965pli0202_1

Baumeister, R. F., Stillwell, A. M., & Heatherton, T. F. (1994). Guilt: An interpersonal approach. *Psychological Bulletin, 115*(2), 243. https://doi.org/10.1037/0033-2909.115.243

Berridge, K. C., Robinson, T. E., & Aldridge, J. W. (2009). Dissecting components of reward: 'liking','wanting', and learning. *Current Opinion in Pharmacology, 9*(1), 65–73. https://doi.org/10.1016/j.coph.2008.12.014

Bible, H. (2003). *The Holy Bible English standard edition (EVS).* Crossway.

Bonanno, G. A., & Keltner, D. (1997). Facial expressions of emotion and the course of conjugal bereavement. *Journal of Abnormal Psychology, 106*(1), 126–137. https://doi.org/10.1037/0021-843X.106.1.126

Breiter, H. C., Gollub, R. L., Weisskoff, R. M., Kennedy, D. N., Makris, N., Berke, J. D., Goodman, J. M., Kantor, H. L., Gastfriend, D. R., Riorden, J. P., Mathew, R. T., Rosen, B. R., & Hyman, S. E. (1997). Acute effects of cocaine on human brain activity and emotion. *Neuron, 19*, 591–611. https://doi.org/10.1016/S0896-6273(00)80374-8

Buss, A. H. (1980). *Self-consciousness and social anxiety.* W. H. Freeman.

Calhoun, L. G., & Tedeschi, R. G. (Eds.) (2014). *Handbook of posttraumatic growth: Research and practice*. Routledge.

Clark, M. T. (2000). *An aquinas reader: Selections from the writings of Thomas aquinas*. Fordham University Press.

Colby, A., & Damon, W. (1992). *Some do care: Contemporary lives of moral commitment*. Free Press.

Csikszentmihalyi, M. (1993). *The evolving self: A psychology for the third millennium*. Harpercollins.

Csikszentmihalyi, M. (2004). Materialism and the evolution of consciousness. In T. Kasser & A. D. Kanner (Eds.), *Psychology and consumer culture: The struggle for a good life in a materialistic world* (pp. 91–106). American Psychological Association.

Curtis, V., Aunger, R., & Rabie, T. (2004). Evidence that disgust evolved to protect from risk of disease. *Proceedings of the Royal Society of London. Series B. Biological Sciences, 271*, 131–133. https://doi.org/10.1098/rsbl.2003.0144

Curtis, V. A., & Biran, A. (2001). Dirt, disgust, and disease: Is hygiene in our genes? *Perspectives in Biology and Medicine, 44*, 17–31. https://doi.org/10.1353/pbm.2001.0001

Damasio, A. R. (2005). *Descartes' error: Emotion, reason, and the human brain*. Penguin.

Damasio, A. R., Grabowski, T. J., Bechara, A., Damásio, H., Ponto, L. L., Parvizi, J., & Hichwa, R. D. (2000). Subcortical and cortical brain activity during the feeling of self-generated emotions. *Nature Neuroscience, 3*(10), 1049–1056. https://doi.org/10.1038/79871

Darwin, C. R. (1965). *The expression of the emotions in man and animals*. University of Chicago Press. (Original work published in 1872.)

Davidson, R. J., & Irwin, W. (1999). The functional neuroanatomy of emotion and affective style. *Trends in Cognitive Sciences, 3*(1), 11–21. https://doi.org/10.1016/S1364-6613(98)01265-0

Deci, E. L., & Ryan, R. M. (1985). *Intrinsic motivation and self-determination in human behavior*. Springer.

Diener, E., & Tay, L. (2017). A scientific review of the remarkable benefits of happiness for successful and healthy living. In *Happiness: Transforming the development landscape* (pp. 90–117). The Centre for Bhutan Studies and GNH.

Durkheim, E. (1951). *Suicide: A study in sociology* (J. A. Spaulding & G. Simpson, Trans.). The Free Press. (Original work published in 1897).

Edelmann, R. J. (1981). Embarrassment: The state of research. *Current Psychological Reviews, 1*, 125–138. https://doi.org/10.1007/bf02979260

Ehrhardt, J. J., Saris, W. E., & Veenhoven, R. (2000). Stability of life-satisfaction over time: Analysis of change in ranks in a national population. *Journal of Happiness Studies, 1*, 177–205. https://doi.org/10.1023/a:1010084410679

Einstein, A. (1954). *Ideas and opinions*. Three Rivers Press.

Eisenberg, N., Fabes, R. A., Miller, P. A., Fultz, J., Shell, R., Mathy, R. M., & Reno, R. R. (1989). Relation of sympathy and personal distress to prosocial behavior: A multimethod study. *Journal of Personality and Social Psychology, 57*(1), 55. https://doi.org/10.1037/0022-3514.57.1.55

Eisenberg, N., Miller, P. A., Shell, R., Mcnalley, S., & Shea, C. (1991). Prosocial development in adolescence: A longitudinal study. *Developmental Psychology, 27*(5), 849. https://doi.org/10.1037/0012-1649.27.5.849

Ekman, P. (1993). Are there basic emoticons? *Psychological Review, 99*(3), 550–553. https://doi.org/10.1037/0033-295X.99.3.550

Ekman, P., & Cordaro, D. (2011). What is meant by calling emotions basic. *Emotion Review, 3*(4), 364–370. https://doi.org/10.1177/1754073911410740

Ekman, P., & Davidson, R. J. (1994). *The nature of emotion: Fundamental questions.* Oxford University Press.

Emmons, R. A., & Mccullough, M. E. (2003). Counting blessings versus burdens: An experimental investigation of gratitude and subjective well-being in daily life. *Journal of Personality and Social Psychology, 84*(2), 377–389.

Emmons, R. A., & McCullough, M. E. (Eds.). (2004). *The psychology of gratitude.* Oxford University Press.

Englander, Z. A., Haidt, J., & Morris, J. P. (2012). Neural basis of moral elevation demonstrated through inter-subject synchronization of cortical activity during free-viewing. *PLoS One, 7*(6), e39384, 1–8. https://doi.org/10.1371/journal.pone.0039384

Festinger, L. (1957). *A theory of cognitive dissonance.* Row, Peterson.

Festinger, L. A. (1954). A theory of social comparison processes. *Human Relations, 7,* 117–140. https://doi.org/10.1177/001872675400700202

Fischer, A., & Giner-Sorolla, R. (2016). Contempt: Derogating others while keeping calm. *Emotion Review, 8*(4), 346–357. https://doi.org/10.1177/1754073915610439

Fischer, N. (1992). *Hybris: A study in the values of honour and shame in ancient Greece.* Aris & Phillips.

Foster, G. M. (1972). The anatomy of envy: A study in symbolic behavior. *Current Anthropology, 13,* 165–202. https://doi.org/10.1086/201267

Fouragnan, E., Retzler, C., & Philiastides, M. G. (2018). Separate neural representations of prediction error valence and surprise: Evidence from an fMRI meta-analysis. *Human Brain Mapping, 39*(7), 2887–2906. https://doi.org/10.1002/hbm.24047

Fox, G. R., Kaplan, J., Damasio, H., & Damasio, A. (2015). Neural correlates of gratitude. *Frontiers in Psychology, 6,* 1491, 1–11. https://doi.org/10.3389/fpsyg.2015.01491

Fredrickson, B. L. (2013). Positive emotions broaden and build. In P. Devine & A. Plant (Eds.), *Advances in experimental social psychology* (vol. 47, pp.1–53). Academic Press. https://doi.org/10.1016/B978-0-12-407236-7.00001-2

Freud, S. (1961). Civilization and its discontents. In *The standard edition of the complete psychological works of sigmund freud* (J. Strachey, Trans., vols 4–5, p. 128). Norton. (Original work published in 1930.)

Frey, B. S., & Stutzer, A. (2018). *Economics of happiness.* Springer International Publishing.

Frijda, N. H. (1999). Emotions and hedonic experience. In D. Kahneman, E. Diener, & N. Schwarz (Eds.), *Well-being: The foundations of hedonic psychology* (pp. 190–210). Russell Sage Foundation.

Frijda, N. H. (2007). *The laws of emotion.* Lawrence Erlbaum.

Fulwiler, C. E., King, J. A., & Zhang, N. (2012). Amygdala-orbitofrontal resting state functional connectivity is associated with trait anger. *Neuroreport, 23*(10), 606–610. https://doi.org/10.1097/WNR.0b013e3283551cfc

Gervais, M., & Wilson, D. S. (2005). The evolution and functions of laughter and humor: A synthetic approach. *The Quarterly Review of Biology, 80*(4), 395–430. https://doi.org/10.1086/498281

Goffman, E. (2017). On face-work. *Interaction Ritual,* 5–46. https://doi.org/10.4324/9780203788387-2

Goleman, D. (2004). *Destructive emotions: How can we overcome them? A scientific dialogue with the Dalai Lama.* Bantam.

Gottman, J. M. (2008). Gottman method couple therapy. In A. S. Gurman (Ed.), *Clinical handbook of couple therapy* (vol. 4, no. 8, pp. 138–164). The Guilford Press.

Gottman, J. M., & Levenson, R. W. (2000). The timing of divorce: Predicting when a couple will divorce over a 14-year period. *Journal of Marriage and Family, 62*(3), 737–745. https://doi.org/10.1111/j.1741-3737.2000.00737.x

Gottman, J. M., & Silver, N. (2015). *The seven principles for making marriage work: A practical guide from the country's foremost relationship expert*. Harmony.

Gouldner, A. W. (1960). The norm of reciprocity: A preliminary statement. *American Sociological Review, 25*, 161–178. https://doi.org/10.2307/2092623

Grant, H., & Higgins, E. T. (2003). Optimism, promotion pride, and prevention pride as predictors of quality of life. *Personality and Social Psychology Bulletin, 29*(12), 1521–1532. https://doi.org/10.1177/0146167203256919

Green, S., Moll, J., Deakin, J. F. W., Hulleman, J., & Zahn, R. (2013). Proneness to decreased negative emoticons in major depressive disorder when blaming others rather than oneself. *Psychopathology, 46*, 34–44. https://doi.org/10.1159/000338632

Green, S., Ralph, M. A. L., Moll, J., Deakin, J. F. W., & Zahn, R. (2012). Guilt-selective functional disconnection of anterior temporal and subgenual cortices in major depressive disorder. *Archives of General Psychiatry, 69*, 1014–1021. https://doi.org/10.1001/archgenpsychiatry.2012.135

Guha, M. (2014). Diagnostic and Statistical Manual of Mental Disorders: DSM-5 (5th edition). *Reference Reviews, 28*(3), 36–37. https://doi.org/10.1108/RR-10-2013-0256

Haidt, J. (2003a). The moral emotions. In R. J. Davidson, K. R. Scherer, & H. H. Goldsmith (Eds.), *Handbook of affective sciences* (vol. 11, pp. 852–870). Oxford University Press.

Haidt, J. (2003b). Elevation and the positive psychology of morality. In C. L. M. Keyes & J. Haidt (Eds.), *Flourishing: Positive psychology and the life well-lived* (pp. 275–289). American Psychological Association.

Haidt, J., Algoe, S., Meijer, Z., Tam, A., & Chandler, E. C. (2002). *Elevation: An emotion that makes people want to do good deeds*. Unpublished Manuscript, University of Virginia.

Harmon-Jones, E., & Mills, J. (2019). An introduction to cognitive dissonance theory and an overview of current perspectives on the theory. In E. Harmon-Jones (Ed.), *Cognitive dissonance: Reexamining a pivotal theory in psychology* (pp. 3–24). American Psychological Association. https://doi.org/10.1037/0000135-001

Hoffman. M. L. (1982). Development of prosocial motivation: Empathy and guilt. In N. Eisenberg (Ed.), *The Development of prosocial behavior* (pp. 218–231). Academic Press. https://doi.org/10.1037/0022-3514.84.2.377

Immordino-Yang, M. H., McColl, A., Damasio, H., & Damasio, A. (2009). Neural correlates of admiration and compassion. *Proceedings of the National Academy of Sciences, 106*(19), 8021–8026. https://doi.org/10.1073/pnas.0810363106

Izard, C. (1977). *Human emotions*. Plenum Press.

Izard, C. E. (1993). Four systems for emotion activation: Cognitive and noncognitive processes. *Psychological Review, 100*(1), 68–90. https://doi.org/10.1037/0033-295X.100.1.68

Kahneman, D. (1999). Objective happiness. In D. Kahneman, E. Diener, & N. Schwarz (Eds.), *Well-being: The foundations of hedonic psychology* (pp. 3–25). Russell Sage Foundation.

Kahneman, D. (2011). *Thinking, fast and slow*. Farrar, Straus, and Giroux.

Kelley, H. H., & Michela, J. L. (1980). Attribution theory and research. *Annual Review of Psychology, 31*(1), 457–501. https://doi.org/10.1146/annurev.ps.31.020180.002325

Klimecki, O. M., Leiberg, S., Lamm, C., & Singer, T. (2013). Functional neural plasticity and associated changes in positive affect after compassion training. *Cerebral Cortex, 23*(7), 1552–1561. https://doi.org/10.1093/cercor/bhs142

Knutson, B., Adams, C. M., Fong, G. W., & Hommer, D. (2001). Anticipation of increasing monetary reward selectively recruits nucleus accumbens. *Journal of Neuroscience, 21*(16), 1–5. https://doi.org/10.1523/JNEUROSCI.21-16-j0002.2001

Koepp, M. J., Gunn, R. N., Lawrence, A. D., Cunningham, V. J., Dagher, A., Jones, T., Brooks, D. J., Bench, C. J., & Grasby, P. M. (1998). Evidence for striatal dopamine release during a video game. *Nature 393*, 266–268. https://doi.org/10.1038/30498

Lama, D., & Cutler, H. (2003). *The art of happiness at work*. Penguin Group.
Lama, D., & Ekman, P. (2008). *Emotionalawareness: Overcoming the obstacles to psychological balance and compassion: A conversation between the Dalai Lama and Paul Ekman*. Times Books.
Lane, R. D., Reiman, E. M., Ahern, G. L., Schwartz, G. E., & Davidson, R. J. (1997). Neuroanatomical correlates of happiness, sadness, and disgust. *The American Journal of Psychiatry, 154*(7), 926–933. https://doi.org/10.1176/ajp.154.7.926
Lazarus, R. S. (1991). *Emotion and adaptation*. Oxford University Press.
Leary, M. R., Tate, E. B., Adams, C. E., Batts Allen, A., & Hancock, J. (2007). Self-compassion and reactions to unpleasant self-relevant events: The implications of treating oneself kindly. *Journal of Personality and Social Psychology, 92*(5), 887. https://doi.org/10.1037/0022-3514.63.2.234
Leget, C., & Olthuis, G. (2007). Compassion as a basis for ethics in medical education. *Journal of Medical Ethics, 33*(10), 617–620. https://dx.doi.org/10.1136/jme.2006.017772
Lewis, C. S. (1955). *Surprised by joy: The shape of my early life*. Harcourt.
Lewis, H. B. (1971). Shame and guilt in neurosis. *Psychoanalytic Review, 58*(3), 419.
Lewis, M. (1975). Infants' social perception: A constructivist view. *Infant Perception: From Sensation to Cognition, 2*, 101–148. https://doi.org/10.1016/b978-0-12-178602-1.50010-5
Lewis, M. (1992). *Shame: The exposed self*. Free Press.
Lord, C. (Ed.). (2013). *Aristotle's politics*. University of Chicago Press.
Lyubomirsky, S. (2008). *The how of happiness*. Penguin Press.
Lyubomirsky, S., King, L., & Diener, E. (2005). The benefits of frequent positive affect: Does happiness lead to success? *Psychological Bulletin, 131*(6), 803. https://doi.org/10.1037/0033-2909.131.6.803
MacBeth, A., & Gumley, A. (2012). Exploring compassion: A meta-analysis of the association between self-compassion and psychopathology. *Clinical Psychology Review, 32*(6), 545–552. https://doi.org/10.1016/j.cpr.2012.06.003
Mathewes, C. (2015). Toward a theology of joy. In M. Volf & J. E. Crisp (Eds.), *Joy and human flourishing: Essays on theology, culture, and the good life* (pp. 63–96). Fortress Press.
McCullough, M. E., Emmons, R. A., & Tsang, J. (2002). The grateful disposition: A conceptual and empirical topography. *Journal of Personality and Social Psychology, 82*, 112–127. https://doi.org/10.1037/0022-3514.82.1.112
McCullough, M. E., Kilpatrick, S., Emmons, R. A., & Larson, D. (2001). Is gratitude a moral affect? *Psychological Bulletin, 127*, 249–266. https://doi.org/10.1037/0033-2909.127.2.249
Miller, R. S. (1996). *Embarrassment: Poise and peril in everyday life*. Guilford Press.
Moll, J., de Oliveira-Souza, R., Eslinger, P. J., Bramati, I. E., Mourão-Miranda, J., Andreiuolo, P. A., & Pessoa, L. (2002). The neural correlates of moral sensitivity: A functional magnetic resonance imaging investigation of basic and moral emotions. *Journal of Neuroscience, 22*(7), 2730–2736. https://doi.org/10.1523/JNEUROSCI.22-07-02730.2002
Moll, J., de Oliveira-Souza, R., Moll, F. T., Ignácio, F. A., Bramati, I. E., Caparelli-Dáquer, E. M., & Eslinger, P. J. (2005). The moral affiliations of disgust: A functional MRI study. *Cognitive and Behavioral Neurology, 18*(1), 68–78. https://doi.org/10.1097/01.wnn.0000152236.46475.a7
Mussweiler, T. (2003). Comparison processes in social judgment: Mechanisms and consequences. *Psychological Review, 110*, 472–489. https://doi.org/10.1037/0033-295x.110.3.472

Neff, K. D., & Vonk, R. (2009). Self-compassion versus global self-esteem: Two different ways of relating to oneself. *Journal of Personality, 77*(1), 23–50. https://doi.org/10.1111/j.1467-6494.2008.00537.x

Niemiec, R. M. (2017). *Character strenghts interventions: A field guide for practitioners*. Hogrefe Publishing.

Oatley, K., & Johnson-Laird, P. N. (1996). The communicative theory of emotions: Empirical tests, mental models, and implications for social interaction. In L. L. Martin & A. Tesser (Eds.), *Striving and feeling: Interactions among goals, affect, and emotion*. Erlbaum.

Ochsner, K. N., Bunge, S. A., Gross, J. J., & Gabrieli, J. D. (2002). Rethinking feelings: An fMRI study of the cognitive regulation of emotion. *Journal of Cognitive Neuroscience, 14*(8), 1215–1229. https://doi.org/10.1162/089892902760807212

Ortony, A., Clore, G. L., &. Collins, A. (1988). *The cognitive structure of emotions*. Cambridge University Press.

Parrott, W. G. (1991). Experiences of envy and jealousy. In P. Salovey (Ed.), *The psychology of jealousy and envy* (pp. 3–30). The Guilford Press.

Parrott, W. G., & Smith, R. H. (1993). Distinguishing the experiences of envy and jealousy. *Journal of Personality and Social Psychology, 64*(6), 906. https://doi.org/10.1037/0022-3514.64.6.906

Patel, S., Pelletier-Bui, A., Smith, S., Roberts, M. B., Kilgannon, H., Trzeciak, S., & Roberts, B. W. (2019). Curricula for empathy and compassion training in medical education: A systematic review. *PLoS One, 14*(8), e0221412. https://doi.org/10.1371/journal.pone.0221412

Patrick, B. C., Hisley, J., & Kempler, T. (2000). What's everybody so excited about? The effects of teacher enthusiasm on student intrinsic motivation and vitality. *Journal of Experimental Education, 68*, 217–236. https://doi.org/10.1080/00220970009600093

Peterson, A. (2016). *Compassion and education: Cultivating compassionate children, schools and communities*. Springer.

Peterson, C., & Seligman, M. E. (2004). *Character strengths and virtues: A handbook and classification*. Oxford University Press.

Phan, K. L., Wager, T., Taylor, S. F., & Liberzon, I. (2002). Functional neuroanatomy of emotion: A meta-analysis of emotion activation studies in PET and fMRI. *Neuroimage, 16*(2), 331–348. https://doi.org/10.1006/nimg.2002.1087

Phillips, M. L., Bullmore, E. T., Howard, R., Woodruff, P. W., Wright, I. C., Williams, S. C. R., Simmons, A., Andrew, C., Brammer, M., & David, A. S. (1998). Investigation of facial recognition memory and happy and sad facial expression perception: An fMRI study. *Psychiatry Research: Neuroimaging, 83*(3), 127–138. https://doi.org/10.1016/S09254927(98)00036-5

Portmann, J. (2000). *When bad things happen to other people*. Routledge.

Rauch, S. L., Shin, L. M., Dougherty, D. D., Alpert, N. M., Orr, S. P., Lasko, M., Macklin, M. L., Fischman, A. J., & Pitman, R. K. (1999). Neural activation during sexual and competitive arousal in healthy men. *Psychiatry Research: Neuroimaging, 91*(1), 1–10. https://doi.org/10.1016/S0925-4927(99)00020-7

Robinson, T. E., & Berridge, K. C. (1993). The neural basis of drug craving: An incentive-sensitization theory of addiction. *Brain Research Reviews, 18*(3), 247–291. https://doi.org/10.1016/0165-0173(93)90013-P

Rolls, E. T. (1999). *The brain and emotion*. Oxford University Press.

Rozin, P., Lowery, L., & Ebert, R. (1994). Varieties of disgust faces and the structure of disgust. *Journal of Personality and Social Psychology, 66*(5), 870. https://doi.org/10.1037/0022-3514.66.5.870

Russell, B. (1930). *The conquest of happiness*. Leverlight.

Ryan, R. M., & Deci, E. L. (2000). Self-determination theory and the facilitation of intrinsic motivation, social development, and well-being. *American Psychologist, 55*(1), 68–78. https://doi.org/10.1037/0003-066x.55.1.68

Salovey, P., & Rodin, J. (1991). Provoking jealousy and envy: Domain relevance and self-esteem threat. *Journal of Social and Clinical Psychology, 10*, 395–413. https://doi.org/10.1521/jscp.1991.10.4.395

Sambataro, F., Dimalta, S., Di Giorgio, A., Taurisano, P., Blasi, G., Scarabino, T., Giannatempo, M. N., & Bertolino, A. (2006). Preferential responses in amygdala and insula during presentation of facial contempt and disgust. *European Journal of Neuroscience, 24*(8), 2355–2362. https://doi.org/10.1111/j.1460-9568.2006.05120.x

Schoeck, H. (1969). *Envy: A Theory of Social Behavior*. Harcourt, Brace and World.

Schoenberg, P. L., Hepark, S., Kan, C. C., Barendregt, H. P., Buitelaar, J. K., & Speckens, A. E. (2014). Effects of mindfulness-based cognitive therapy on neurophysiological correlates of performance monitoring in adult attention-deficit/hyperactivity disorder. *Clinical Neurophysiology, 125*(7), 1407–1416. https://doi.org/10.1016/j.clinph.2013.11.031

Schwarz, N. (1990). Feelings as information: Informational and motivational functions of affective states. In E. T. Higgins & R. M. Sorrentino (Eds.), *Handbook of motivation and cognition: Foundations of Social Behavior* (vol. 2, pp. 527–561). Guilford Press.

Schwarz, N. (1998). Warmer and more social: Recent developments in cognitive social psychology. *Annual Review of Sociology, 24*, 239–264. https://doi.org/10.1146/annurev.soc.24.1.239

Simmel, G. (1950). *The sociology of Georg Simmel*. Free Press.

Simon-Thomas, E. R., Godzik, J., Castle, E., Antonenko, O., Ponz, A., Kogan, A., & Keltner, D. J. (2012). An fMRI study of caring vs. self-focus during induced compassion and pride. *Social Cognitive and Affective Neuroscience, 7*(6), 635–648. https://doi.org/10.1093/scan/nsr045

Singer, T., & Ricard, M. (2015). *Caring economics: Conversations on altruism and compassion, between scientists, economists, and the Dalai Lama*. Picador.

Smith, A. (1976). *The theory of moral sentiments* (6th ed.). Clarendon Press. (Original work published in 1790.)

Smith, A. (2010). *The wealth of nations: An inquiry into the nature and causes of the wealth of nations*. Harriman House Limited.

Smith, R. H., Diener, E., & Wedell, D. H. (1989). Intrapersonal and social comparison determinants of happiness: A range-frequency analysis. *Journal of Personality and Social Psychology, 56*, 317–325. https://doi.org/10.1037/0022-3514.56.3.317

Spencer, H. (1855). *Principles of Psychology*. Liberty Fund Inc.

Spitzer, M., Fischbacher, U., Herrnberger, B., Grön, G., & Fehr, E. (2007). The neural signature of social norm compliance. *Neuron, 56*(1), 185–196. https://doi.org/10.1016/j.neuron.2007.09.011

Sprengelmeyer, R., Rausch, M., Eysel, U. T., & Przuntek, H. (1998). Neural structures associated with recognition of facial expressions of basic emotions. *Proceedings of the Royal Society of London. Series B: Biological Sciences, 265*(1409), 1927–1931. http://dx.doi.org/10.1098/rspb.1998.0522

Stevenson, R. J., & Repacholi, B. M. (2005). Does the source of an interpersonal odour affect disgust? A disease risk model and its alternatives. *European Journal of Social Psychology, 35*, 375–401. https://doi.org/10.1002/ejsp.263

Storbeck, J., & Clore, G. (2005). With sadness comes accuracy; With happiness, false memory: Mood and the false memory effect. *Psychological Science, 16*(10), 785–791. https://doi.org/10.1111/j.1467-9280.2005.01615.x

Suardi, A., Sotgiu, I., Costa, T., Cauda, F., & Rusconi, M. (2016). The neural correlates of happiness: A review of PET and fMRI studies using autobiographical recall methods. *Cognitive, Affective, & Behavioral Neuroscience, 16*(3), 383–392. https://doi.org/10.3758/s13415-016-0414-7

Takahashi, H., Kato, M., Matsuura, M., Mobbs, D., Suhara, T., & Okubo, Y. (2009). When your gain is my pain and your pain is my gain: Neural correlates of envy and schadenfreude. *Science, 323*(5916), 937–939. https://doi.org/10.1126/science.1165604

Tangney, J. P., Miller, R. S., Flicker, L., & Barlow, D. H. (1996). Are shame, guilt, and embarrassment distinct emotions? *Journal of Personality and Social Psychology, 70,* 1256–1269. https://doi.org/10.1037/0022-3514.70.6.1256

Tedeschi, R. G., & Calhoun, L. G. (2004). Posttraumatic growth: Conceptual foundations and empirical evidence. *Psychological Inquiry, 15*(1), 1–18. https://doi.org/10.1207/s15327965pli1501_01

Tesser, A., Gatewood, R., & Driver, M. (1968). Some determinants of gratitude. *Journal of Personality and Social Psychology, 9,* 233–236. *https://doi.org/10.1037/h0025905*

Thorne, F. C. (1959). The etiology of sociopathic reactions. *American Journal of Psychotherapy, 13*(2), 319–330. https://doi.org/10.1176/appi.psychotherapy.1959.13.2.319

Thrash, T. M., & Elliot, A. J. (2004). Inspiration: Core characteristics, component processes, antecedents, and function. *Journal of Personality and Social Psychology, 87*(6), 957–973. https://doi.org/10.1037/0022-3514.87.6.957

Tracy, J. L., & Robins, R. W. (2004). Putting the self into self-conscious emotions: A theoretical model. *Psychological Inquiry, 15,* 103–125. https://doi.org/10.1207/s15327965pli1502_01

Tracy, J. L., & Robins, R. W. (2007). The psychological structure of pride: A tale of two facets. *Journal of Personality and Social Psychology, 92*(3), 506–525. https://doi.org/10.1037/0022-3514.92.3.506

Tusche, A., Böckler, A., Kanske, P., Trautwein, F. M., & Singer, T. (2016). Decoding the charitable brain: Empathy, perspective taking, and attention shifts differentially predict altruistic giving. *Journal of Neuroscience, 36*(17), 4719–4732. https://doi.org/10.1523/JNEUROSCI.3392-15.2016

Vaillant, G. E. (2008). *Spiritual evolution: How we are wired for faith, hope, and love.* Broadway Books.

Van de Ven, N., Zeelenberg, M., & Pieters, R. (2009). Leveling up and down: The experiences of benign and malicious envy. *Emotion, 9*(3), 419–429. https://doi.org/10.1037/a0015669

Verplanken, B., & Holland, R. W. (2002). Motivated decision making: Effects of activation and self-centrality of values on choices and behavior. *Journal of Personality and Social Psychology, 82*(3), 434–447. https://doi.org/10.1037/0022-3514.82.3.434

Volf, M. (2015). The crown of the good life: A hypothesis. In M. Volf & J. E. Crisp (Eds.), *Joy and human flourishing: Essays on theology, culture, and the good life* (pp. 127–136). Fortress Press.

Vytal, K., & Hamann, S. (2010). Neuroimaging support for discrete neural correlates of basic emotions: A voxel-based meta-analysis. *Journal of Cognitive Neuroscience, 22*(12), 2864–2885. https://doi.org/10.1162/jocn.2009.21366

Wagner, U., N'Diaye, K., Ethofer, T., & Vuilleumier, P. (2011). Guilt-specific processing in the prefrontal cortex. *Cerebral Cortex, 21*(11), 2461–2470. https://doi.org/10.1093/cercor/bhr016

Wang, S., Xu, X., Zhou, M., Chen, T., Yang, X., Chen, G., & Gong, Q. (2017). Hope and the brain: Trait hope mediates the protective role of medial orbitofrontal cortex spontaneous activity against anxiety. *Neuroimage, 157,* 439–447. https://doi.org/10.1016/j.neuroimage.2017.05.056

Watkins, P. C., Emmons, R. A., Greaves, M. R., & Bell, J. (2018). Joy is a distinct positive emotion: Assessment of joy and relationship to gratitude and well-being. *The*

Journal of Positive Psychology, 13(5), 522–539. https://doi.org/10.1080/17439760.2017.1414298

Waugh, C. E., & Fredrickson, B. L. (2006). Nice to know you: Positive emotions, self-other overlap, and complex understanding in the formation of new relationships. *Journal of Positive Psychology, 1*, 93–106. https://doi.org/10.1080/17439760500510569

Westerink, H. (2009). *A dark trace: Sigmund freud on the sense of guilt* (Vol. 8). Leuven University Press.

Williams, L. A., & Desteno, D. (2008). Pride and perseverance: The motivational role of pride. *Journal of Personality and Social Psychology, 94*(6), 1007–1017. https://doi.org/10.1037/0022-3514.94.6.1007

Williamson, P. B. (2009). *General's patton principles for life and leadership*. Managements & Systems Consultants (MSC), Inc.

Yu, H., Gao, X., Zhou, Y., & Zhou, X. (2018). Decomposing gratitude: Representation and integration of cognitive antecedents of gratitude in the brain. *Journal of Neuroscience, 38*(21), 4886–4898. https://doi.org/10.1523/JNEUROSCI.2944-17.2018

Zahn, R., de Oliveira-Souza, R., Bramati, I., Garrido, G., & Moll, J. (2009). Subgenual cingulate activity reflects individual differences in empathic concern. *Neuroscience Letters, 457*(2), 107–110. https://doi.org/10.1016/j.neulet.2009.03.090

Chapter 2

The Broaden-and-Build Theory of Positive Emotions

By the end of the 1990s, most scientific studies focused only on negative emotions. The cause for this was the fact that the field of psychology was intended to help solve the psychological problems of individuals, particularly after World War II (Seligman & Csikszentmihalyi, 2000).

Scientists thought that all emotions had the characteristic of triggering trends for action (Frijda et al., 1989). This is true for negative emotions. For example, fear generates the behavioral tendency to escape or paralyze, anger generates the behavioral tendency to assault, disgust generates the trend to move away from the object of disgust, and so on (Frijda et al., 1989). From an evolutionary point of view, negative emotions triggered action tendencies, especially to increase the probability of survival (Darwin, 1872/1965).

In this way, fear represents a crucial emotion that stimulates prey to a rapid escape; anger is important in the sense that the predator undertakes the utmost effort to catch and fight with their prey or to dominate the territory in the face of other male competitors, as in the case of chimpanzees; disgust triggers the action of avoiding ingesting or being in contact with something that can cause disease (Darwin, 1872/1965).

Another important idea is that the physiological changes generated by negative emotions lead to specific repertoires of behavior. Thus, the action comes accompanied by a set of physiological changes that facilitate specific action repertoires and are compatible with their function.

For example, fear activates all body stress in a brain system called the "flight-or-fight system," and so does anger, disgust and sadness. The problem is that these actions triggered by emotions do not apply to positive emotions. For example, joy does not trigger any specific predisposition to an action. On the contrary, for example, joy generates only the propensity to explore new things (Fredrickson, 2013).

Professor Barbara Fredrickson was the first researcher to scientifically demonstrate that certain emotions do not have specific trends for action (Fredrickson & Levenson, 1998). There is another fundamental feature that differs from negative emotions. Positive emotions tend to alleviate the negative effects of

DOI: 10.4324/9781003503880-3

negative emotions (Fredrickson & Levenson, 1998). One study proved that the stimulus to positive emotions tended to accelerate recovery from a negative physiological arousal caused by negative emotions, compared to the stimulus to the emotion of sadness and a neutral stimulus (Fredrickson & Levenson, 1998). It was an incredible scientific discovery, because it paved the way for a different function of positive emotions compared to negative emotions.

For example, one study showed that the presence of positive emotions tends to accelerate cardiovascular recovery generated by negative emotions, i.e., a person who has positive emotions as a resource tends to recover more rapidly from the physical effects of negative emotions compared to a person who does not generate positive emotions (Fredrickson & Levenson, 1998).

This is one of the aspects that has a fundamental implication and application in the regulation of emotions, because, in addition to people having the means to regulate negative emotions, positive emotions can be a powerful resource to help reduce the negative emotions and turn them into positive emotions.

A neuroscientific study revealed the relationship between variability in the heart rate in the face of stimuli of various emotions and activation of the MPFC (Lane et al., 2009). Another study demonstrated that the activation of the prefrontal cortex during a stressful situation predicted faster recovery from stress as well as a lower presence of negative emotions and the presence of positive emotions in a stressful situation (Yang et al., 2018)

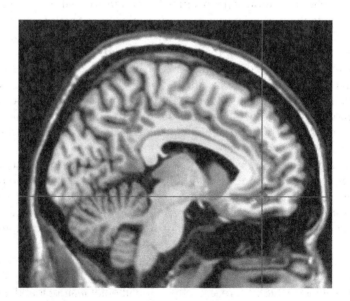

Figure 2.1 Representation of the ventromedial prefrontal cortex, at the intersection of the horizontal line with the vertical line.

This study corroborates the pioneering idea of Professor Antonio Damasio, who reported that patients with lesions in this region tend to lose relevant emotional and motivational aspects (Damasio, 2005). Some studies that have already become classic in neuroscience have demonstrated a brain asymmetry between positive and negative emotions.

These studies revealed that positive emotions tend to activate more often the right prefrontal cortex (Davidson, 1992, 1998, 2004a, 2004b). Research has also shown that those who have a resilient and positive affective emotion tend to generate positive emotions more regularly (associated with increased activation of the left prefrontal cortex), are able to recover more quickly from stressful events and negative emotions and are able to regulate the activation of the amygdala, which is associated with the intensity of emotional arousal (Davidson, 2004a).

However, cerebral asymmetry does not seem to be due only to positive and negative emotions, what the researchers usually call emotional valence, but it is because of more complex aspects (Davidson, 2004b). Researchers, rather than thinking that this asymmetry is related only to the difference between positive and negative emotions, proposed the idea that the left prefrontal cortex is more connected to approach motivation, and the right prefrontal cortex is more associated with avoidance motivation (Reuter-Lorenz & Davidson, 1981; Harmon-Jones, 2004).

Classically, the motivation for action (approach motivation) has been associated with motivation toward a positive stimulus, particularly in the direction of positive external goals (Lang & Bradley, 2008). However, recent research challenges this idea, because it was found that anger is also activated in the left prefrontal cortex, and these new data change the concept of approach motivation as a push to "move on," regardless of whether the emotion is positive or negative, or the motivation comes from an external or inner stimulus (Harmon-Jones et al., 2013).

Thus, both positive emotions and negative anger are associated with motivation for action, which helps explain why this emotion also relates to activation of the left prefrontal cortex in conjunction with positive emotions (Harmon-Jones, 2004). Regarding the motivation for avoidance, whose goal is for the individual to position themselves in a defensive manner and to avoid action, it is more related to the negative emotions of sadness and the activation of the cortex (Davidson, 1998, 1992).

This way, when we talk about emotional intelligence, it is important to understand the motivations underlying emotions, because it is possible to approach only at the level of emotion, when, in fact, the origin of emotion is associated with motivational tendencies – in this case, in motivation for action versus motivation for avoidance (Lang & Bradley, 2008; Harmon-Jones et al., 2013).

Based on the idea that positive emotions have a set of functions different from negative emotions, Professor Barbara Fredrickson proposed a theory for the expansion and construction of positive emotions – the broaden-and-build theory of positive emotions (Fredrickson & Levenson, 1998; Fredrickson, 2001). The broaden-and-build theory of positive emotions states that positive emotions tend to broaden the mental and behavioral repertoire of action and to generate actions that build important resources for individuals (Fredrickson & Levenson, 1998; Fredrickson, 2001).

This theory proposes two hypotheses: the hypothesis of enlargement and the construction hypothesis.

The Broaden Hypothesis

The broaden hypothesis suggests that positive emotions tend to expand or broaden the cognitive and mental capacity of human beings in various fields as well as the repertoire of behaviors and actions.

There are several scientific studies that confirm this hypothesis. The first basic evidence of the broaden hypothesis is that positive emotions tend to broaden the spectrum of visual attention, with a significant increase in vision (Rowe et al., 2007).

Another study demonstrated that while negative emotions tend to reduce focus on a specific target, positive emotions tend to broaden the focus of attention (Rowe et al., 2007; Fredrickson & Branigan, 2005).

Most of the evidence for the broaden hypothesis comes from several studies conducted by a pioneer in the scientific approach to positive emotions, Professor Alice Isen, in the 1980s and 1990s (Isen, 1990). In those studies, it was demonstrated that positive emotions tend to increase inclusion in cognitive categorization exercises (Isen & Daubman, 1984) and many other cognitive domains, such as improving creative problem-solving skills (Isen et al., 1987), making more flexible and efficient decision-making strategies (Isen & Means, 1983), increasing cognitive creativity (Isen et al., 1985), increasing openness to new information in decision-making processes (Estrada et al., 1997), developing creativity and innovation at work (Amabile et al., 2005), improving efficiency in solving clinical problems in a medical context (Isen et al., 1991), increasing self-regulation (Aspinwall, 1998) and increasing happiness levels (Diener et al., 2009).

In addition, one study found that positive emotions increase the predisposition for a greater number of behavioral initiatives (Fredrickson & Branigan, 2005) and increase the inclusion of other individuals in the sense of their own self, that is, see others as closer to the self (Waugh & Fredrickson, 2006).

It has also been found that positive emotions tend to increase the inclusion of other social groups in the sense of self, in the context of different social groups, decreasing the so-called distance between "us and them" and consequently decreasing levels of racism (Dovidio et al., 1995).

Another study demonstrated that positive emotions tend to get people to expand their circle of trusted people (Dunn & Schweitzer, 2005). It was also shown that emotions tend to lead people to develop empathy and compassion for people of different races and cultures (Nelson, 2009).

In summary, and to facilitate the reader's understanding, positive emotions tend to broaden the focus of attention, improve problem-solving skills more creatively, make decision-making strategies more flexible and efficient, increase cognitive creativity, increase creativity at work and, consequently, organizational innovation, improve efficiency in solving clinical problems in medical decision-making, develop emotional self-regulation, increase the predisposition to more behavioral activities, increase the inclusion of others in the self, increase the inclusion of other social groups and reduce racism, increase happiness, grow the circle of trusted people, increase empathy and compassion for people of different races and cultures.

The Build Hypothesis

The construction hypothesis states that positive emotions can lead to actions that build lasting personal resources and stimulate a positive spiral of growth. These resources generated by positive emotions are cognitive, psychological, social and physical.

The main cognitive resources identified scientifically (Fredrickson et al., 2008) in people and caused by positive emotions were the following:

Mindfulness – A revealing scientific study identified that the effects of meditation on mindfulness in reducing stress are mediated by the increase in positive reassessment caused by positive emotions (Garland et al., 2011). Thus, it seems that positive emotions tend to increase positive emotional reassessment, and this can reduce stress levels (Garland et al., 2011). This means that this relationship has two paths: emotions mediate the effect of meditation on mindfulness in stress reduction, and mindfulness meditation tends to increase positive emotions (Garland et al., 2011). This two-way relationship is scientifically designated as a positive spiral (Garland et al., 2011).

Development of various alternatives of thinking and decision-making strategies – As mentioned earlier, positive emotions tend to improve problem-solving capacity more effectively (Isen et al., 1987), make decision-making strategies more flexible and efficient (Isen & Means, 1983) and increase openness to new information in decision-making processes (Estrada et al., 1997).

Ability to appreciate the present – This is one of the strategies developed scientifically to increase people's emotional well-being. The capacity to appreciate is the predisposition to focus and enjoy past, present and future positive events (Bryant, 1989). This trait, which varies from person to person, is seen as a kind of perceived control over positive emotions, where the pleasure from the events comes through conscious will (Bryant, 1989). There are three components in the act of appreciating: the anticipation of positive events, the reminiscence of positive events that have occurred in the past and the appreciation of the moment and the present (Bryant, 2003).

Ability to appreciate and anticipate the future – A scientific study has shown that positive emotions tend to increase people's ability to appreciate both the present and the anticipation of a positive event (Fredrickson et al., 2008). The cognitive resource of appreciating the present and the future is related to higher levels of happiness and satisfaction with life and lower levels of depression and negative emotions (Bryant & Veroff, 2007; Hurley & Kwon, 2012).

The main psychological resources identified by the investigation by Fredrickson et al. (2008) as results of positive emotions are as follows:

Resilience – This is the ability to successfully overcome adversity and to adapt adequately to difficult circumstances (Masten, 2007). Some scientific studies have shown that positive emotions tend to increase resilience to adversity (Fredrickson et al., 2008, 2003).
In the latter study, it was demonstrated that those who most often generated positive emotions of gratitude, interest and love before the attack of September 11, 2001, tended to deal with the crisis generated after the tragedy more positively, to develop fewer symptoms of depression and to exhibit greater resilience compared to the control group (Fredrickson et al., 2003). This study provided strong scientific evidence of the power of positive emotions as a resource for resilience.
In two scientific studies, it was found that positive emotions contributed, partly for the ability to regulate emotions efficiently in adverse situations, which has been demonstrated by cardiovascular recovery from a negative event (Tugade & Fredrickson, 2004). In a third study, these researchers found that positive emotions play an important role in the ability to create a positive meaning in the face of adverse circumstances (Tugade & Fredrickson, 2004).
Other important evidence showed that positive emotions help people deal with daily stress and grief (Ong et al., 2006). Yet another study found that positive emotions tend to increase the ability to find benefits in adversity (Hart et al., 2008).
One of the most important characteristics of positive emotions is their undoing effect of the harmful impact of negative emotions and events. It has already been found that some people smile in the face of adversity, while others do not. Researchers have found that people who smile during a sad film demonstrate greater cardiovascular recovery after the movie compared to people who do not smile during the adversity or negative stimulus (Fredrickson & Levenson, 1998).

Self-acceptance – This is one of the most important capabilities for the evolution of the real self of any human being. Without self-acceptance, individual progress is severely compromised, because the individual does not allow themselves to reflect seriously and profoundly about the negative aspects of their real self that need to be corrected and refined

throughout life, in the direction of evolution and complexity. The first major psychologist to emphasize the importance of self-acceptance was Carl Rogers, in his monumental book *On Becoming a Person*, originally written in 1961 (Rogers, 1961).
- Moral elevation, admiration and compassion are important positive emotions that stimulate self-acceptance, therefore, its importance for the evolution of the self. Without these positive emotions, it would be difficult for the human being to be open to accepting the interior changes necessary for its evolution. For example, moral elevation generates the genuine desire of the person to seek to be better as a human being from a moral point of view, and this strongly stimulates self-acceptance. One of the biggest obstacles to self-acceptance is ego-based pride and vanity (Tracy & Robins, 2007). Pride and vanity generate a rejection of any negative information relating to the self. In this regard, emotional intelligence is important to help regulate the excess of pride; to help generate moral elevation, compassion and admiration; and to encourage a predisposition for self-acceptance. This resource is also considered a fundamental intrinsic value for the progress of the self and tends to increase levels of positive emotional well-being and happiness and decrease levels of anxiety and depression (Kasser et al., 2014).
- *Learned optimism* – The main investigator that developed the concept of learned optimism was Professor Martin Seligman, one of the fathers of positive psychology. According to him, optimists tend to believe that negative events are more temporary than permanent and rapidly recover from failures, while pessimists take longer to recover (Seligman, 2006). Optimists also believe that positive events happen due to permanent factors, rather than thinking of them as temporary. Research has shown that optimists are more successful and generally healthier (Seligman, 2006).
- *Clear purpose of life goals* – Research shows that when someone has clear life goals, they tend to be more successful and have higher levels of happiness and well-being (Ciani et al., 2011). What research shows is that positive emotions generate emotional resources that stimulate the definition of a clearer purpose of life (Fredrickson et al., 2008).

The main social resources that positive emotions help build are as follows (Fredrickson et al., 2008):

Social support given to others – There are many scientific studies that demonstrate that positive emotions tend to increase helping behavior (Batson & Shaw, 1991; Darwin, 1872/1965; Davis, 1994; de Waal, 2008; Frank, 1988; Hume, 1777/1960; Loewenstein & Small, 2007; Slovic, 2007; Smith, 1790/1976). For example, one study showed that those who had a background training in compassion tended to be more selfless and to help others more compared to those who felt no compassion (Weng et al., 2013).

Two scientific studies found that one who observes someone doing good, and feels the moral emotion of elevation, tends to be more selfless (Schnall et al., 2010). This study demonstrated that moral elevation makes people more selfless when compared to the positive emotion of fun (Schnall et al., 2010).

In a longitudinal study of 258 secondary school teachers, researchers found that positive emotions at work, coming from flow state (Csikszentmihalyi, 1990) and intrinsic motivation (Ryan & Deci, 2000; Deci & Ryan, 1985), tended to generate personal resources of self-effectiveness and organizational resources, such as increased social support for employees and business objectives, in a spiral dynamic (Salanova et al., 2006).

Another study showed that people who are carriers of a higher emotional positivity often experience a higher frequency of positive emotions daily, through activities on how to help others, interact constructively with others, play, learn and devote themselves to spiritual activities (Catalino & Fredrickson, 2011).

The data suggest that these positive emotions promote a positive spiral, which in turn increases the motivation for positive emotional activities (Catalino & Fredrickson, 2011). This study demonstrates that positive emotions drive resources to help others, spreading a positive spiral composed of positive emotions – positive activities, including helping others, lead to mindfulness and positive emotions (Catalino & Fredrickson, 2011).

Perception of received social support – An experimental study demonstrated that the increase in positive emotions tends to develop people's perception who receive social support from others (Fredrickson et al., 2008). This study suggests that when we feel positive emotions, we have the perception that we are more socially supported by others compared to when we are feeling neutral or negative emotions (Fredrickson et al., 2008).

Positive relationships with others – The previously mentioned study demonstrated that people who feel positive emotions more often are those who claim to have more positive relationships with others, also in a positive spiral: positive emotions lead to positive relationships lead to positive emotions, and so on (Fredrickson et al., 2008).

The main physical resources generated by positive emotions are as follows:

Decreased symptoms of disease – One of the most amazing aspects of the effects of positive emotions are their positive effects on physical health, called physical resources, such as reducing depression; reducing the severity of flu symptoms; decreasing mortality of patients; reducing symptoms of hypertension, diabetes and infections of the respiratory tract; and increasing longevity.

Depression – The study already mentioned that stimulating positive emotions through loving-kindness meditation demonstrated that positive emotions tended to decrease the symptoms of depression as well as to build cognitive, psychological and social resources, and these also had the tendency to decrease the symptoms of depression (Fredrickson et al., 2008).

Increased vagal tone in the parasympathetic nervous system – Another biological consequence of the effects of positive emotions on physical health is the vagal tone. This is rated as a slight arrhythmia, although functional (Grossman, 1983). The vagal tone refers to the activity of the vagus nerve, a fundamental component of the parasympathetic nervous system. This branch of the nervous system is not under conscious control and is largely responsible for regulating various body parts when at rest. You can analyze the activity of the vagal tone in various situations, such as heart rate reduction, vasodilation and the constriction of vessels; the glandular activity in the heart, lungs and digestive tract; as well as control of gastrointestinal sensitivity, motility and inflammation. The vagal tone is typically analyzed in the context of cardiac function, but it is also useful in evaluation of emotional regulation and other processes that change or are altered by changes in the activity of the parasympathetic nervous system. The vagal tone is considered a key measure of the influence of the parasympathetic system in the heart, not only because it has been related to positive emotional traits (Oveis et al., 2009) but also for their strong links with physical and mental health (Porges, 2007; Thayer & Sternberg, 2006). The vagal tone functions as a personal physical resource with implications for cardiovascular and metabolic functioning as well as for emotional and social well-being.

One study showed that the increase in vagal tone predicted a growth of positive emotions within 2 months, and conversely, the increase in positive emotions predicted the increase of vagal tone, in a positive spiral (Kok & Fredrickson, 2010). By using the vagal tone as a biological marker, data from this study suggest that people may benefit from their health through the frequency of positive emotions (Kok & Fredrickson, 2010). The limitation of this study is that it was not able to demonstrate a causal relationship between positive emotions and vagal tone but only a positive association.

To overcome this limitation, these researchers performed another study that demonstrated a causal relationship between positive emotions and the increase in vagal tone (Kok et al., 2013). This was one of the few studies that demonstrated a causal relationship between positive emotions and health, due to the biological measure of the cardiac vagal tone. Most studies reveal only a positive association, not a cause-and-effect relationship.

Decreased flu symptoms – For example, a study showed positive emotions support the effect of IL-6 proteins on expression of the symptoms of rhinovirus infection (Doyle et al., 2006). This means that this study showed that people who generate positive emotions more often tend to have fewer symptoms of rhinovirus infection compared to those who have negative emotions more regularly (Doyle et al., 2006).

Decrease in the mortality of patients with AIDS – Another study demonstrated that patients with AIDS who generate positive emotions more frequently tend to have a lower mortality rate (Moskowitz, 2003).

Hypertension, diabetes and respiratory tract infections – Another study carried out by researchers from the Harvard Medical School demonstrated that high levels of hope tend to decrease the symptoms of hypertension, diabetes and respiratory tract infections (Richman et al., 2005). In addition, researchers revealed that high levels of curiosity decrease symptoms of hypertension and diabetes (Richman et al., 2005). Researchers concluded that positive emotions are powerful features of health prevention (Richman et al., 2005).

Increased longevity – Some studies have shown that those who generate positive emotions more often tend to live longer (Danner et al., 2001; Moskowitz, 2003; Ostir et al., 2000). A meta-analysis of almost 300 scientific studies concluded that emotions produce success and health as well as reflect these good results (Lyubomirsky et al., 2005). A study that followed 180 Catholic nuns showed that those who wrote positive emotional content in their diaries were strongly associated with a greater longevity 60 years later (Danner et al., 2001).

A team of researchers from the Texas School of Medicine on a project funded by the National Institutes of Health demonstrated, in a sample of 2282 participants aged between 65 and 99 years, that a high frequency of positive emotions had 50 percent more probability of surviving after 2 years compared to the other groups composed of participants with neutral or negative emotions (Ostir et al., 2000).

In another meta-analysis that reviewed scientific studies on the relationship between happiness and health, in 62 of them, which involved approximately 1,250,000 participants, it was found, with statistical rigor, that happiness is a "protector" factor of mortality (Martín-María et al., 2017). One of the parameters of happiness is emotional well-being, in which happiness is measured by a higher frequency of positive emotions compared to negative emotions (Steptoe et al., 2015).

A scientific way to measure the biological age of human beings is through telomeres. These are protein complexes contained in the terminal part of eukaryotic chromosomes that protect DNA from instability and cell degradation (Blackburn, 2005). In general, telomeres tend to decrease their length with cell division (i.e., by aging) but can also be lengthened with the enzyme telomerase (Blackburn, 2005). Shorter telomeres predict a higher incidence of risk of mortality and age-related diseases, including cardiovascular diseases and diabetes (Blackburn, 2005).

Some studies suggest that shorter telomeres are associated with psychosocial factors (Puterman & Epel, 2012; Shalev et al., 2013) as well as higher levels of stress, objectively measured as life adversities (Tyrka et al., 2010) and subjectively perceived as stress (Mathur et al., 2016). Shorter telomeres are also related to higher levels of anxiety (Hoen et al., 2013) and depression (Schutte & Malouff, 2015).

On the other hand, conscientiousness in female children is related to longer telomeres 40 years later (Edmonds et al., 2015).

Conscientiousness is the personality trait characterized by a predisposition to work harder, to adhere to social norms, with a tendency to have the capacity to plan and control impulses as well as perform a task properly and take seriously the obligations toward others (Jackson et al., 2010).

One study demonstrated a positive relationship between high levels of optimism and emotional intelligence, and a longer length of telomeres, compared with lower levels of optimism and emotional intelligence (Schutte et al., 2016). Although this study demonstrated a positive association between emotional intelligence and a longer telomere length, this relationship showed no scientific evidence in terms of cause-and-effect relationship.

In a study that aimed to find a causal relationship between loving-kindness meditation and increased telomere length, 145 middle-aged participants were randomly placed in three groups: a control group (without meditation), another group that performed the mindfulness meditation and a third group that performed loving-kindness meditation for 6 weeks (Le Nguyen et al., 2019). The length of telomeres in participants was measured before and after the study. It was shown that within 3 weeks after the study (which lasted 6 weeks), there was a decrease in the length of the telomeres of the control group and of the group that carried out the mindfulness meditation, which is a biological marker of aging, but not in the group that performed loving-kindness meditation (Le Nguyen et al., 2019).

These data suggest that participants who performed loving-kindness meditation have not aged from a biological point of view like the other two groups (Le Nguyen et al., 2019). And, as mentioned earlier, this same experimental paradigm of loving-kindness meditation demonstrated, with a relationship of cause and effect, a significant increase in positive emotions and in all resources built on positive emotions, including physical resources, such as vagal tone (Fredrickson et al., 2008; Kok et al., 2013).

The scientific evidence described in the broaden-and-build theory of positive emotions demonstrates the importance of positive emotions in the lives of human beings. The research also suggests the importance of frequently and systematically stimulating positive emotions in human beings, so that everyone, and society in general, can benefit from this wonderful resource. To do so, educational organizations must learn effective ways to stimulate frequent emotions in the context of families, businesses and all other human organizations.

References

Amabile, T. M., Barsade, S. G., Mueller, J. S., & Staw, B. M. (2005). Affect and creativity at work. *Administrative Science Quarterly, 50*(3), 367–403. https://doi.org/10.2189/asqu.2005.50.3.367

Aspinwall, L. G. (1998). Rethinking the role of positive affect in self-regulation. *Motivation and Emotion, 22*(1), 1–32. https://doi.org/10.1023/a:1023080224401

Batson, C. D., & Shaw, L. L. (1991). Evidence for altruism: Toward a pluralism of prosocial motives. *Psychological Inquiry, 2*(2), 107–122. https://doi.org/10.1207/s15327965pli0202_1

Blackburn, E. H. (2005). Telomeres and telomerase: Their mechanisms of action and the effects of altering their functions. *FEBS Letters, 579*, 859–862. https://doi.org/10.1016/j.febslet.2004.11.036

Bryant, F. B. (1989). A four-factor model of perceived control: Avoiding, coping, obtaining, and savoring. *Journal of Personality, 57*, 773–797. https://doi.org/10.1111/j.1467-6494.1989.tb00494.x

Bryant, F. B. (2003). Savoring beliefs inventory (SBI): A scale for measuring beliefs about savouring. *Journal of Mental Health, 12*, 175–196. https://doi.org/10.1080/0963823031000103489

Bryant, F. B., & Veroff, J. (2007). *Savoring: A new model of positive experience*. Lawrence Erlbaum Associates.

Catalino, L. I., & Fredrickson, B. L. (2011). A tuesday in the life of a flourisher: The role of positive emotional reactivity in optimal mental health. *Emotion, 11*(4), 938–950. https://doi.org/10.1037/a0024889

Ciani, K. D., Sheldon, K. M., Hilpert, J. C., & Easter, M. A. (2011). Antecedents and trajectories of achievement goals: A self-determination theory perspective. *British Journal of Educational Psychology, 81*(2), 223–243. https://doi.org/10.1348/000709910X517399

Csikszentmihalyi, M. (1990). *Flow: The psychology of optimal experience*. Harper and Row.

Damasio, A. R. (2005). *Descartes' error: Emotion, reason, and the human brain*. Penguin.

Danner, D. D., Snowdon, D. A., & Friesen, W. V. (2001). Positive emotions in early life and longevity: Findings from the nun study. *Journal of Personality and Social Psychology, 80*(5), 804. https://doi.org/10.1037/0022-3514.80.5.804

Darwin, C. R. (1965). *The Expression of the emotions in man and animals*. University of Chicago Press. (Original work published in 1872).

Davidson, R. J. (1992). Anterior cerebral asymmetry and the nature of emotion. *Brain and Cognition, 20*(1), 125–151. https://doi.org/10.1016/0278-2626(92)90065-T

Davidson, R. J. (1998). Anterior electrophysiological asymmetries, emotion, and depression: Conceptual and methodological conundrums. *Psychophysiology, 35*(5), 607–614. https://doi.org/10.1017/S0048577298000134

Davidson, R. J. (2004a). Well-being and affective style: Neural substrates and biobehavioral correlates. *Philosophical Transactions of the Royal Society of London. Series B: Biological Sciences, 359*(1449), 1395–1411. https://doi.org/10.1098/rstb.2004.1510

Davidson, R. J. (2004b). What does the prefrontal cortex "do" in affect: Perspectives on frontal EEG asymmetry research. *Biological Psychology, 67*(1–2), 219–234. https://doi.org/10.1016/j.biopsycho.2004.03.008

Davis, M. H. (1994). *Empathy: A social psychological approach*. Westview Press.

de Waal, F. B. M. (2008). Putting the altruism back into altruism: The evolution of empathy. *Annual Review of Psychology, 59*, 279–300. https://doi.org/10.1146/annurev.psych.59.103006.093625

Deci, E. L., & Ryan, R. M. (1985). *Intrinsic motivation and self-determination in human behavior*. Springer.

Diener, E., Sandvik, E., & Pavot, W. (2009). Happiness is the frequency, not the intensity, of positive versus negative affect. In F. Strack, M. Argyle, & N. Schwarz (Eds.), *Assessing well-being* (pp. 213–231). Springer.

Dovidio, J., Gaertner, S., Isen, A., Rust, M., & Guerra, P. (1995). Positive affect and the reduction of intergroup bias. In C. Sedikides, J. Schopler, & C. A. Insko (Eds.), *Intergroup cognition and intergroup behavior* (pp. 337–366). Erlbaum.

Doyle, W. J., Gentile, D. A., & Cohen, S. (2006). Emotional style, nasal cytokines, and illness expression after experimental rhinovirus exposure. *Brain, Behavior, and Immunity, 20*, 175–181. https://doi.org/10.1016/j.bbi.2005.05.005

Dunn, J. R., & Schweitzer, M. E. (2005). Feeling and believing: The influence of emotion on trust. *Journal of Personality and Social Psychology, 88*(5), 736–748. https://doi.org/10.1037/0022-3514.88.5.736

Edmonds, G. W., Côté, H. C., & Hampson, S. E. (2015). Childhood conscientiousness and leukocyte telomere length 40 years later in adult women-preliminary findings of a prospective association. *PLoS One, 10*(7), e0134077. https://doi.org/10.1371/journal.pone.0134077

Estrada, C. A., Isen, A. M., & Young, M. J. (1997). Positive affect facilitates integration of information and decreases anchoring in reasoning among physicians. *Organizational Behavior and Human Decision Processes, 72*, 117–135. https://doi.org/10.1006/obhd.1997.2734

Frank, R. H. (1988). *Passions within reason: The strategic role of the emotions*. Norton.

Fredrickson, B. L. (2001). The role of positive emotions in positive psychology: The broaden-and-build theory of positive emotions. *American Psychologist, 56*(3), 218–226. *https://doi.org/10.1037/0003-066x.56.3.218*

Fredrickson, B. L. (2013). Positive emotions broaden and build. In P. Devine & A. Plant (Eds.), *Advances in experimental social psychology,* (Vol. 47, pp. 1–53). Academic Press. https://doi.org/10.1016/B978-0-12-407236-7.00001-2

Fredrickson, B. L., & Branigan, C. (2005). Positive emotions broaden thought-action repertoires: Evidence for the broaden-and-build model. *Cognition and Emotion, 19*, 313–332. https://doi.org/10.1080/02699930441000238

Fredrickson, B. L., Cohn, M. A., Coffey, K. A., Pek, J., & Finkel, S. M. (2008). Open hearts build lives: Positive emotions, induced through loving-kindness meditation, build consequential personal resources. *Journal of Personality and Social Psychology, 95*(5), 1045. https://doi.org/10.1037/a0013262

Fredrickson, B. L., & Levenson, R. W. (1998). Positive emotions speed recovery from the cardiovascular sequelae of negative emotions. *Cognition and Emotion, 12*(2), 191–220. https://doi.org/10.1080/026999398379718

Fredrickson, B. L., Tugade, M. M., Waugh, C. E., & Larkin, G. R. (2003). What good are positive emotions in crisis? A prospective study of resilience and emotions following the terrorist attacks on the United States on September 11th, 2001. *Journal of Personality and Social Psychology, 84*(2), 365–376. https://doi.org/10.1037/0022-3514.84.2.365

Frijda, N. H., Kuipers, P., & Schure, E. (1989). Relations among emotion, appraisal, and emotional action readiness. *Journal of Personality and Social Psychology, 57*, 212–228. https://doi.org/10.1037/0022-3514.57.2.212

Garland, E. L., Gaylord, S. A., & Fredrickson, B. L. (2011). Positive reappraisal mediates the stress-reductive effects of mindfulness: An upward spiral process. *Mindfulness, 2*(1), 59–67. https://doi.org/10.1007/s12671-011-0043-8

Grossman, P. (1983). Respiration, stress, and cardiovascular function. *Psychophysiology, 20*(3), 284–300. https://doi.org/10.1111/j.1469-8986.1983.tb02156.x

Harmon-Jones, E. (2004). Contributions from research on anger and cognitive dissonance to understanding the motivational functions of asymmetrical frontal brain activity. *Biological Psychology, 67*(1–2), 51–76. https://doi.org/10.1016/j.biopsycho.2004.03.003

Harmon-Jones, E., Harmon-Jones, C., & Price, T. F. (2013). What is approach motivation? *Emotion Review, 5*(3), 291–295. https://doi.org/10.1177/1754073913477509

Hart, S. L., Vella, L., & Mohr, D. C. (2008). Relationships among depressive symptoms, benefit-finding, optimism, and positive affect in multiple sclerosis patients after psychotherapy for depression. *Health Psychology, 27*(2), 230–238. https://doi.org/10.1037/0278-6133.27.2.230

Hoen, P. W., Rosmalen, J. G., Schoevers, R. A., Huzen, J., Van Der Harst, P., & De Jonge, P. (2013). Association between anxiety but not depressive disorders and leukocyte telomere length after 2 years of follow-up in a population-based sample. *Psychological Medicine, 43*(4), 689–697. https://doi.org/10.1017/S0033291712001766

Hume, D. (1960). *An enquiry concerning the principles of morals.* Open Court. (Original work published 1777).

Hurley, D. B., & Kwon, P. (2012). Results of a study to increase savoring the moment: Differential impact on positive and negative outcomes. *Journal of Happiness Studies, 13*(4), 579–588. https://doi.org/10.1007/s10902-011-9280-8

Isen, A. M. (1990). The influence of positive and negative affect on cognitive organization: Implications for development. In N. Stein, B. Leventhal, & T. Trabasso (Eds.), *Psychological and biological processes in the development of emotion.* Erlbaum.

Isen, A. M., & Daubman, K. A. (1984). The influence of affect on categorization. *Journal of Personality and Social Psychology, 47*, 1206–1217. https://doi.org/10.1037/0022-3514.47.6.1206

Isen, A. M., Daubman, K. A., & Nowicki, G. P. (1987). Positive affect facilitates creative problem solving. *Journal of Personality and Social Psychology, 52*, 1122–1131. https://doi.org/10.1037/0022-3514.52.6.1122

Isen, A. M., Johnson, M. M., Mertz, E., & Robinson, G. F. (1985). The influence of positive affect on the unusualness of word associations. *Journal of Personality and Social Psychology, 48*(6), 1413. https://doi.org/10.1037/0022-3514.48.6.1413

Isen, A. M., & Means, B. (1983). The influence of positive affect on decision-making strategy. *Social Cognition, 2*, 18–31. https://doi.org/10.1521/soco.1983.2.1.18

Isen, A. M., Rosenzweig, A. S., & Young, M. J. (1991). The influence of positive affect on clinical problem solving. *Medical Decision Making, 11*, 221–227. https://doi.org/10.1177/0272989X9101100313

Jackson, J. J., Wood, D., Bogg, T., Walton, K. E., Harms, P. D., & Roberts, B. W. (2010). What do conscientious people do? Development and validation of the behavioral indicators of conscientiousness (BIC). *Journal of Research in Personality, 44*(4), 501–511. https://doi.org/10.1016/j.jrp.2010.06.005

Kasser, T., Rosenblum, K. L., Sameroff, A. J., Deci, E. L., Niemiec, C. P., Ryan, R. M., Árnadóttir, O., Bond, R., Dittmar, H., Dungan, N., & Hawks, S. (2014). Changes in materialism, changes in psychological well-being: Evidence from three longitudinal studies and an intervention experiment. *Motivation and Emotion, 38*(1), 1–22. https://doi.org/10.1007/s11031-013-9371-4

Kok, B. E., Coffey, K. A., Cohn, M. A., Catalino, L. I., Vacharkulksemsuk, T., Algoe, S. B., Brantley, M., & Fredrickson, B. L. (2013). How positive emotions build physical health: Perceived positive social connections account for the upward spiral between positive emotions and vagal tone. *Psychological Science, 24*(7), 1123–1132. https://doi.org/10.1177/0956797612470827

Kok, B. E., & Fredrickson, B. L. (2010). Upward spirals of the heart: Autonomic flexibility, as indexed by vagal tone, reciprocally and prospectively predicts positive emotions and social connectedness. *Biological Psychology, 85*(3), 432–436. https://doi.org/10.1016/j.biopsycho.2010.09.005

Lane, R. D., McRae, K., Reiman, E. M., Chen, K., Ahern, G. L., & Thayer, J. F. (2009). Neural correlates of heart rate variability during emotion. *Neuroimage, 44*(1), 213–222. https://doi.org/10.1016/j.neuroimage.2008.07.056

Lang, P. J., & Bradley, M. M. (2008). Appetitive and defensive motivation is the substrate of emotion. In A. Elliott (Ed.), *Handbook of approach and avoidance motivation* (pp. 51–66). Psychology Press.

Le Nguyen, K. D., Lin, J., Algoe, S. B., Brantley, M. M., Kim, S. L., Brantley, J., Salzberg, S., & Fredrickson, B. L. (2019). Loving-kindness meditation slows biological aging in novices: Evidence from a 12-week randomized controlled trial. *Psychoneuroendocrinology, 108,* 20–27. https://doi.org/10.1016/j.psyneuen.2019.05.020

Loewenstein, G., & Small, D. A. (2007). The scarecrow and the tin man: The vicissitudes of human sympathy and caring. *Review of General Psychology, 11*(2), 112–126. https://doi.org/10.1037/1089-2680.11.2.112

Lyubomirsky, S., King, L., & Diener, E. (2005). The benefits of frequent positive affect: Does happiness lead to success? *Psychological Bulletin, 131*(6), 803. http://dx.doi.org/10.1037/0033-2909.131.6.803

Martín-María, N., Miret, M., Caballero, F. F., Rico-Uribe, L. A., Steptoe, A., Chatterji, S., & Ayuso-Mateos, J. L. (2017). The impact of subjective well-being on mortality: A meta-analysis of longitudinal studies in the general population. *Psychosomatic Medicine, 79*(5), 565–575. https://doi.org/10.1097/PSY.0000000000000444

Masten, A. S. (2007). Resilience in developing systems: Progress and promise as the fourth wave rises. *Development and Psychopathology, 19*(3), 921–930. https://doi.org/10.1017/S0954579407000442

Mathur, M. B., Epel, E., Kind, S., Desai, M., Parks, C. G., Sandler, D. P., & Khazeni, N. (2016). Perceived stress and telomere length: A systematic review, meta-analysis, and methodologic considerations for advancing the field. *Brain, Behavior, and Immunity, 54,* 158–169. https://doi.org/10.1016/j.bbi.2016.02.002

Moskowitz, J. T. (2003). Positive affect predicts lower risk of AIDS mortality. *Psychosomatic Medicine, 65*(4), 620–626. https://doi.org/10.1097/01.PSY.0000073873.74829.23

Nelson, D. W. (2009). Feeling good and open-minded: The impact of positive affect on cross cultural empathic responding. *The Journal of Positive Psychology, 4*(1), 53–63. https://doi.org/10.1080/17439760802357859

Ong, A. D., Bergeman, C. S., Bisconti, T. L., & Wallace, K. A. (2006). Psychological resilience, positive emotions, and successful adaptation to stress in later life. *Journal of Personality and Social Psychology, 91*(4), 730–749. https://doi.org/10.1037/0022-3514.91.4.730

Ostir, G. V., Markides, K. S., Black, S. A., & Goodwin, J. S. (2000). Emotional well-being predicts subsequent functional independence and survival. *Journal of the American Geriatrics Society, 48*(5), 473–478. https://doi.org/10.1111/j.1532-5415.2000.tb04991.x

Oveis, C., Cohen, A. B., Gruber, J., Shiota, M. N., Haidt, J., & Keltner, D. (2009). Resting respiratory sinus arrhythmia is associated with tonic positive emotionality. *Emotion, 9*(2), 265–270. https://doi.org/10.1037/a0015383

Porges, S. W. (2007). The polyvagal perspective. *Biological Psychology, 74*(2), 116–143. https://doi.org/10.1016/j.biopsycho.2006.06.009

Puterman, E., & Epel, E. (2012). An intricate dance: Life experience, multisystem resiliency, and rate of telomere decline throughout the lifespan. *Social and Personality Psychology Compass, 6*(11), 807–825. https://doi.org/10.1111/j.1751-9004.2012.00465.x

Reuter-Lorenz, P., & Davidson, R. J. (1981). Differential contributions of the two cerebral hemispheres to the perception of happy and sad faces. *Neuropsychologia, 19*(4), 609–613. https://doi.org/10.1016/0028-3932(81)90030-0

Richman, L. S., Kubzansky, L., Maselko, J., Kawachi, I., Choo, P., & Bauer, M. (2005). Positive emotion and health: Going beyond the negative. *Health Psychology, 24*(4), 422. https://doi.org/10.1037/0278-6133.24.4.422

Rogers, Carl. (1961). *On becoming a person: A therapist's view of psychotherapy.* Constable.

Rowe, G., Hirsh, J. B., & Anderson, A. K. (2007). Positive affect increases the breadth of attentional selection. *Proceedings of the National Academy of Sciences, 104*(1), 383–388. https://doi.org/10.1073/pnas.0605198104

Ryan, R. M., & Deci, E. L. (2000). Self-determination theory and the facilitation of intrinsic motivation, social development, and well-being. *American Psychologist, 55*(1), 68–78. https://doi.org/10.1037/0003-066x.55.1.68

Salanova, M., Bakker, A. B., & Llorens, S. (2006). Flow at work: Evidence for an upward spiral of personal and organizational resources. *Journal of Happiness Studies, 7*(1), 1–22. https://doi.org/10.1007/s10902-005-8854-8

Schnall, S., Roper, J., & Fessler, D. M. (2010). Elevation leads to altruistic behavior. *Psychological Science, 21*(3), 315–320. https://doi.org/10.1177/0956797609359882

Schutte, N. S., & Malouff, J. M. (2015). The association between depression and leukocyte telomere length: A meta-analysis. *Depression and Anxiety, 32*(4), 229–238. https://doi.org/10.1002/da.22351

Schutte, N. S., Palanisamy, S. K., & McFarlane, J. R. (2016). The relationship between positive psychological characteristics and longer telomeres. *Psychology & Health, 31*(12), 1466–1480. https://doi.org/10.1080/08870446.2016.1226308

Seligman, M. E. (2006). *Learned optimism: How to change your mind and your life.* Vintage.

Seligman, M. E. P., & Csikszentmihalyi, M. (2000). Positive psychology: An introduction. *American Psychologist, 55*, 5–14. https://doi.org/10.1037/0003-066x.55.1.5

Shalev, I., Entringer, S., Wadhwa, P. D., Wolkowitz, O. M., Puterman, E., Lin, J., & Epel, E. S. (2013). Stress and telomere biology: A lifespan perspective. *Psychoneuroendocrinology, 38*, 1835–1842. https://doi.org/10.1016/j.psyneuen.2013.03.010

Slovic, P. (2007). If I look at the mass I will never act: Psychic numbing and genocide. *Judgment and Decision Making, 2*(2), 79–95. https://doi.org/10.1007/978-90-481-8647-1_3

Smith, A. (1976). *The theory of moral sentiments* (6th ed.). Clarendon Press. (Original work published in 1790).

Steptoe, A., Deaton, A., & Stone, A. A. (2015). Subjective wellbeing, health, and ageing. *The Lancet, 385*(9968), 640–648. https://doi.org/10.1016/S0140-6736(13)61489-0

Thayer, J. F., & Sternberg, E. (2006). Beyond heart rate variability: Vagal regulation of allostatic systems. *Annals of the New York Academy of Sciences, 1088*(1), 361–372. https://doi.org/10.1196/annals.1366.014

Tracy, J. L., & Robins, R. W. (2007). Emerging insights into the nature and function of pride. *Current Directions in Psychological Science, 16*(3), 147–150. https://doi.org/10.1111/j.1467-8721.2007.00493.x

Tugade, M. M., & Fredrickson, B. L. (2004). Resilient individuals use positive emotions to bounce back from negative emotional experiences. *Journal of Personality and Social Psychology, 86*(2), 320–333. https://doi.org/10.1037/0022-3514.86.2.320

Tyrka, A. R., Price, L. H., Kao, H. T., Porton, B., Marsella, S. A., & Carpenter, L. L. (2010). Childhood maltreatment and telomere shortening: Preliminary support for an effect of early stress on cellular aging. *Biological Psychiatry, 67*(6), 531–534. https://doi.org/10.1016/j.biopsych.2009.08.014

Waugh, C. E., & Fredrickson, B. L. (2006). Nice to know you: Positive emotions, self-other overlap, and complex understanding in the formation of new relationships. *Journal of Positive Psychology, 1*, 93–106. https://doi.org/10.1080/17439760500510569

Weng, H. Y., Fox, A. S., Shackman, A. J., Stodola, D. E., Caldwell, J. Z., Olson, M. C., Rogers, G. M., & Davidson, R. J. (2013). Compassion training alters altruism and neural responses to suffering. *Psychological Science, 24*(7), 1171–1180. https://doi.org/10.1177/0956797612469537

Yang, X., Garcia, K. M., Jung, Y., Whitlow, C. T., McRae, K., & Waugh, C. E. (2018). vmPFC activation during a stressor predicts positive emotions during stress recovery. *Social Cognitive and Affective Neuroscience, 13*(3), 256–268. https://doi.org/10.1093/scan/nsy012

Chapter 3

Complexity, Evolution and the Neural Bases of Self

Know Thyself

– Temple at Delphi (Parke & Wormell, 1956)

The Evolution of Self and the Concept of Complexity

Usually, people tend to think that something is complex when it is "complicated." In fact, in biology, an organism can be called complex when it has two fundamental characteristics: differentiation and integration (Csikszentmihalyi, 1993).

The best example to explain the concept of complexity is through the analysis of the human body and its systems: cardiovascular, circulatory, digestive, endocrine, etc. Each system of the human body has functions with a high level of specialization. The human being can be considered more complex compared to an amoeba, for example, because the number of systems with functions and capabilities are larger than that of the amoeba, which has systems with little differentiation or functional specialization.

In addition to a higher level of differentiation, composed of systems with a high level of specialization, when the human organism is healthy, these different systems are regulated in harmony, i.e., they work in an integrated manner, this being the concept of integration (Csikszentmihalyi, 1993). That is why the concept of evolution is associated with the concept of complexity. An organism is complex when it is differentiated and integrated simultaneously, that is, in addition to having many functional differentiations, it works in harmony, i.e., in an integrated way.

Now let us move this concept to the social dimension. Let's reflect on a genius. A genius can be quite differentiated, with a remarkable intellectual capacity and a broad scholarship, but may be someone quite negative from mental and emotional points of view, and difficult in interpersonal relationships, with a dubious morality. It can be said that this person is highly differentiated but is not integrated.

DOI: 10.4324/9781003503880-4

This person is differentiated but not in harmony because they do not produce positive outcomes for themself or others. Although this person is differentiated, they cannot be considered complex or more evolved. According to this example, what will promote complexity and evolution in this person is the ability to learn integration practices, which will both promote intrapersonal harmony and improve the quality of interpersonal relationships. In this case, to achieve complexity in their process of evolution, the person should develop mechanisms for personal and social integration.

Now let's consider another example. Let's imagine someone who lives in a rural area that is very poor. Despite the poverty, the person lives happily with one's own, in a simple and harmonious life, and in harmony and happiness with the community that surrounds them. This person possesses a group of moral virtues acquired by the practical experience of life and creates very positive and constructive outcomes for everyone, living in mental and emotional balance.

It can be said that this person has a high level of integration but not differentiation. Thus, to become complex, they need new elements of differentiation: different experiences, different situations and different learning for the acquisition of new skills, knowledge and virtues. This person is going to need these new experiences to evolve toward complexity.

Although this person has harmony and integration, to evolve, they need differentiation, to develop new knowledge, experiences and virtues. While in the first example the individual must learn elements of integration and harmony, in the second, the individual needs differentiation. Every human being needs to develop elements of differentiation and integration, and each case should be analyzed individually, with the aim of evolution of the self. A society composed of educational organizations dedicated to helping all people to develop the self, both in relationship with themselves (intrapersonal) and in relationship with others (interpersonal), makes a major contribution to the evolution of human beings.

But what is the use of knowing yourself? It is not possible to evolve the self without knowing deeply our real self. But before we get to know ourselves, we need motivation and direction to this self. It is not enough to know ourselves well without knowing which direction we want to follow during our evolution. The end or goal of evolution is given by the ideal and the values that each one cultivates and occupies in our minds as a focus of attention. The process is how the evolution of the self will occur (Csikszentmihalyi, 1993).

The process of evolution of animals has occurred with the sole purpose of ensuring survival and reproduction for the preservation of species (Darwin, 1964). Of course, in human beings, the same occurs, but in this case, the issue is more complex, because there are other differentiating factors to be integrated than merely survival.

Because human beings have a huge qualitative difference in relation to other animals, which is self-awareness, abstract mental concepts internalized to the sense of self are crucial for human evolution toward complexity, such as motivations, values and purpose of life, not just as a factor of survival, like other animals.

Toward human beings, survival becomes not only an end by itself but a means for self-development (Csikszentmihalyi, 1993).

Of course, within a hierarchy of needs, survival will always be at the base of the pyramid, but as humans are much more complex, there are many other important motivations, when survival is already guaranteed (Maslow, 1954). According to Maslow, the motivation for self-realization is the most important intrinsic type of motivation of human beings (Maslow, 1954).

While animals instinctively and automatically develop their complexity through the struggle for survival, due to the phenomenon of self-awareness and self-reflection, we also have the ability to reflect on our own development and develop a path of evolution of the self in a planned and intentional way (Csikszentmihalyi, 1993). We already have the tendency to do this but still in a rudimentary and unconscious way.

We are pushed by the social forces of culture where we are born, and we are strongly influenced by our family and the environment around us. Research in experimental social psychology has shown that our self is much more shaped and influenced by culture and genes than what we normally know (Csikszentmihalyi, 1993). That is, therefore, the importance of rescuing the value of Classical Greek thought connected to the idea of knowing thyself. It is based on seeking knowledge of the real self, in the sense of creating a society focused on the evolution of the individual self, without illusions of ego or social illusions (Csikszentmihalyi, 1993).

Classical Greece was the cradle of our Western civilization, but fundamental principles have been lost in many ways, such as the idea of paideia, which is nothing more than a concept whose objective was education from the inside out, i.e., focused on the development of the potential of the self of each one of us. According to Socrates, the development of the self could only happen safely if we know in depth our real self, so we would not fall into illusions about ourselves.

According to the paideia, achieving wisdom in a safe and reliable way, that is, without illusions, requires us to start the journey of self-discovery from the identification of our own ignorance, that is, the identification of our real self. The identification of one's own ignorance, using it to deepen the knowledge of our real selves (composed of virtues and weaknesses), is the starting point through which wisdom will be built without illusions.

Socrates created a maieutic method, composed of techniques of questioning in order to find the truth, and according to the philosopher, wisdom is only possible when one accurately discovers their own ignorance about the things of life and about oneself. By being able to figure out what are the things we ignore about ourselves, we will be able to identify our real self, and from there, we will be able to start to develop our own self without illusions.

But for this to happen, the Greeks created the concept of ideal (Jaeger, 1985), which is, as mentioned earlier, the direction, the end to achieve, and the journey of learning toward the ideal. Before you begin the conscious journey of the evolution of the real self, it is necessary to know which direction to take. And the ideals are the direction, that is, the ends to achieve; the evolution of the real self is the means, the process of development.

Therefore, it is necessary to create a reference of the ideal self, its end and its direction, and this developmental process is part of the in-depth analysis of the real self, so that this journey can be done realistically and safely in the system of gradual realization of the self (concept of self-realization). The greater is the development of the real self, as a natural reaction, the happier human beings will be.

For the Greeks, the evolution of the real self toward evolution and complexity, composed of differentiation and integration, will ensure a higher level of happiness and well-being.

In this phase of ideas about the evolution of self, the reader, eager for practical recommendations regarding how to start the process of evolution of the self, tends to pose the following questions: How do we start the evolutionary process? What are the practical steps?

Regarding these practical issues, we suggest the reader see Part II of this book, particularly particularly the detailed explanation of Professor Richard Boyatzis's intentional change theory and self-directed learning (Boyatzis, 2002).

The Neural Bases of Self

The neural bases of self are selectively associated with the MPFC (medial frontal gyrus) and dorsomedial prefrontal cortex (superior frontal gyrus) as well as the left temporal cortex (Ochsner et al., 2004).

In a previous study, this region of the MPFC was also related to a neural network called a "default mode network" (Gusnard et al., 2001). According to some studies, this default mode network is activated in the moment when the individual is inside the fMRI machine without actively performing any cognitive task. It is as if when the individual is not thinking about anything or with no focus on any specific task (this state is called "resting state"), this default mode network is automatically activated. According to some researchers, one of the main regions of this network is the MPFC, which is related to the evaluation processing of the self, even if unconsciously (Gusnard et al., 2001).

However, the default mode network is not restricted to the MPFC. These regions are deactivated as soon as the individual initiates a conscious cognitive task and activated when the individual is in a "resting state," reflecting an intrinsic self-activity, even if unconsciously (Raichle, 2015).

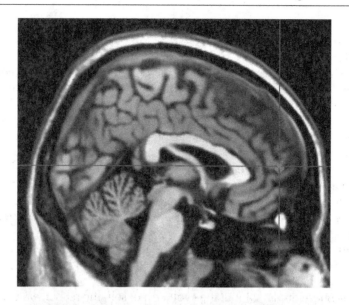

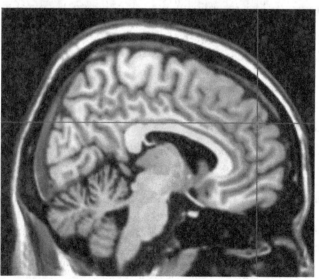

Figure 3.1 Representation of (a) the medial prefrontal cortex and (b) the dorsomedial prefrontal cortex, at the intersection of the horizontal line with the vertical line.

This activation of the self's perspective is contrasted with the perspective of others, since the perspective of others activated the dorsolateral prefrontal cortex, or inferior frontal gyrus (Ochsner et al., 2004).

Complexity, Evolution and the Neural Bases of Self 79

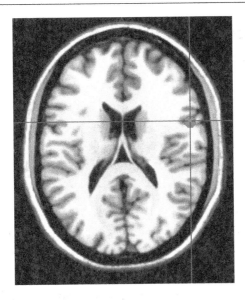

Figure 3.2 Representation of the dorsolateral prefrontal cortex, at the intersection of the horizontal line with the vertical line.

References

Boyatzis, R. E. (2002). Unleashing the power of self-directed learning. In R. Sims (Ed.), *Changing the way we manage change: The consultants speak*. Quorum Books.

Csikszentmihalyi, M. (1993). *The evolving self: A psychology for the third millennium*. Harpercollins.

Darwin, C. (1964). *On the origin of species: A facsimile of the first edition*. Harvard University Press.

Gusnard, D. A., Akbudak, E., Shulman, G. L., & Raichle, M. E. (2001). Medial prefrontal cortex and self-referential mental activity: Relation to a default mode of brain function. *Proceedings of the National Academy of Sciences, 98*(7), 4259–4264. https://doi.org/10.1073/pnas.071043098

Jaeger, W. (1985). *Early christianity and greek paideia*. Harvard University Press.

Maslow, A. H. (1954). *Motivation and personality*. Harper and Row.

Ochsner, K. N., Knierim, K., Ludlow, D. H., Hanelin, J., Ramachandran, T., Glover, G., & Mackey, S. C. (2004). Reflecting upon feelings: An fMRI study of neural systems supporting the attribution of emotion to self and other. *Journal of Cognitive Neuroscience, 16*(10), 1746–1772. https://doi.org/10.1162/0898929042947829

Parke, H. W., & D. Wormell, D. (1956). *The delphic oracle, vol. 1: The history* (p. 389). Blackwell.

Raichle, M. E. (2015). The brain's default mode network. *Annual Review of Neuroscience, 38*, 433–447. https://doi.org/10.1146/annurev-neuro-071013-014030

Chapter 4

The Brain Mechanisms of Emotional Self-Awareness

Before describing the neural bases of emotional self-awareness, it is important to understand as best as possible its concept and its importance for emotional intelligence.

Emotional self-awareness is the ability to accurately recognize the emotion or emotions felt at a certain moment, their intensity, and how that emotion (or emotions) affects one's own thoughts and behaviors, as well as the ability to understand how thoughts, values and motivations affect emotions (Mayer & Salovey, 1997).

In addition, it can be considered the mental reflection on an emotional impulse, about its causes and consequences, and the planning of appropriate responses in a certain context are also capabilities that can be encompassed as emotional self-awareness, and which precede emotional self-control.

Emotional self-awareness is the most critical capacity of emotional intelligence, without which its development would never be possible. After all, how can we learn to manage an emotion if we do not have self-awareness of what emotion we are feeling? A large part of emotional intelligence involves thinking intelligently about emotional impulses, their causes and consequences, and about the appropriate emotional response in a specific situation. This set of capabilities can be encompassed in emotional self-awareness.

Emotional intelligence assumes, as the name implies, that intelligence (i.e., mind, reason, and cognition) has the ability to regulate emotions. If this principle did not exist, there could not exist the concept of emotional intelligence.

In this way, emotional self-awareness should be the first practical training for learning emotional intelligence. And since we were not brought up to investigate ourselves and to become aware of the emotions we are feeling, it is normal that, at first, we see that emotional self-awareness is a more complex exercise than it seems cognitively. Emotional self-awareness requires systematic training of self-observation and reflection on one's own emotions, and how they affect our thoughts and behaviors, or how our thoughts affect our emotions.

Another important aspect of emotional intelligence is the fact that without self-awareness, there is no possibility of emotional self-control. We can only

DOI: 10.4324/9781003503880-5

develop strategies for self-control of emotions identified through emotional self-awareness on a first stage. That is why people find it so difficult to develop emotional self-control, because it requires emotional self-awareness as a capacity that precedes emotional self-control.

The Role of the Anterior Insula in Emotional Self-Awareness

The anterior insular cortex, or simply anterior insula, contains representations of visceral states of the body, which provides the basis for all subjective emotions, and possibly emotional self-awareness, in agreement with James-Lange's theory of emotions and the somatic marker hypothesis of Professor Antonio Damasio (Damasio, 2005; Craig, 2002).

Using the positron emission tomography (PET) diagnostic technique, a study showed that cold temperatures objectively felt by the body are represented in the posterior insula, while the subjective evaluations of these stimuli are related to the activation of the medial insula and much more markedly in the anterior insular cortex, suggesting a passage of insula posterior to the medial and from this to the anterior insula as the stimulus integrates a more complex amount of interoceptive information (visceral sensations) and subjective and abstract information (Craig, 2003).

Another study based on fMRI demonstrated that the self-awareness of the heartbeat is correlated with the activation of the anterior insula (Critchley et al., 2004). It is known, according to another study, that self-consciousness of the heartbeat is an interoceptive measure that establishes with an individual subjective emotional self-awareness (Barret et al., 2004), which confirms the hypothesis of the passage of activation of the insula to the anterior insula, as self-awareness incorporates more complex and subjective information about a stimulus.

It has already been scientifically demonstrated that the anterior insula is closely related to both the visceral perceptions of the body (interoception) and the subjective experiences (Zaki & Ochsner, 2012). It is also suggested that the anterior insula, from an evolutionary point of view, may have progressed from self-awareness of the states of the body to emotional self-awareness (Damasio, 1999).

Thus, based on hundreds of neuroscientific studies, the anterior insula is suggested to be the brain region most related to emotional self-awareness (Craig, 2009).

An interesting finding is the fact that in most neuroscientific studies, when the anterior insula was activated associated with emotional self-awareness, the anterior cingulate cortex was also activated (Craig, 2009). All emotion has its content and a set of behavioral trends associated with the content of this same emotion, called action tendencies derived from emotions (Frijda et al., 1989).

82 What Neuroscience Can Teach Us About Emotional Intelligence

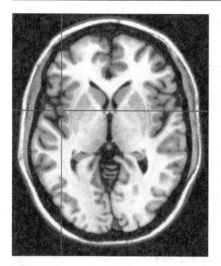 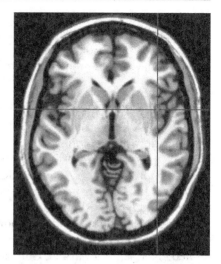

Figure 4.1 Representation of (a) the right anterior insula and (b) the left anterior insula associated with emotional self-awareness, at the intersection of the horizontal line with the vertical line.

A respected neuroscientist proposed that while the anterior insula is associated with the content of emotion and its emotional self-awareness, the cingulate cortex relates to the behavioral tendency linked to this emotion (Craig, 2009).

Based on a series of neuroscientific studies, it has been suggested that the anterior insula represents emotional self-awareness, and that the anterior cingulate cortex represents control of directed effort. It is as if the anterior insula, which is the neural structure related to emotional self-awareness, transmits the information of emotion and its emotional salience to the anterior cingulate cortex region, and this would help to think, reflect and mentally plan which behavioral action tendency is most appropriate for a given situation. The neural structure that is more associated with this behavioral tendency is the dorsal anterior cingulate cortex (Craig, 2009; Satpute et al., 2013).

The Role of the Medial Orbitofrontal Cortex as a Gateway to Self-Awareness

In 2003, Schore proposed that the mOFC could be characterized as "the gateway to awareness." This researcher suggested the idea that in this brain region, the interoceptive (i.e., coming from the interior) and exteroceptive (i.e., coming from the outside) stimuli may occur pre-consciously and may not be a conscious process (Schore, 2003). The proposal that this region is the gateway to awareness is according to previous studies (Kawasaki et al., 2001).

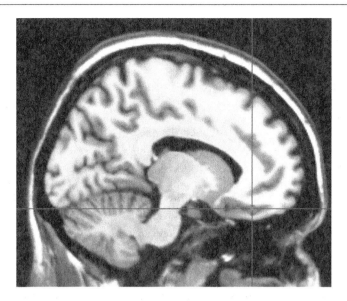

Figure 4.2 Representation of the orbitofrontal cortex associated with the "gateway to self-awareness," at the intersection of the horizontal line with the vertical line.

Activation in this region was observed as fast as in 200 ms after emotional stimulus (Kawasaki et al., 2001). In contrast, the lateral prefrontal cortex was activated only 300 to 800 ms after the emotional stimulation (Kawasaki et al., 2001). These data may imply that the emotional stimulus related to the self can be initially represented in the mOFC before being processed in other cortical regions (Northoff & Bermpohl, 2004). Thus, it is possible to predict that injuries in this region may compromise the processing of representations (Northoff & Bermpohl, 2004).

The Role of the Anterior Cingulate Cortex in Emotional Self-Awareness

It is important to detail a little more about this brain region and its role in emotional self-consciousness. In the anterior cingulate cortex, there are three fundamental functions (Bush et al., 2000): It has specific modules for processing sensory, cognitive and emotional information; it integrates information from a variety of sources (e.g., motivation, evaluation of errors and representations from cognitive neural networks and emotional); and it influences the activities of other regions, such as cognitive, motor, endocrine and visceral.

The researchers distinguished three processes underlying emotional self-evaluation (Satpute et al., 2013): sensitivity to emotional intensity, emotional self-awareness, and self-awareness about the states of the body arising from activation of the anterior insula and amygdala; categorization of emotional experience through the ability to name and to verbalize the emotion – emotional literacy (associated with the VLPFC); and mental reasoning and reflections associated with the emotional experience (associated with the dorsomedial prefrontal cortex and anterior cingulate cortex).

The anterior insula, as already mentioned, is related to self-awareness, and self-awareness about the states of the body (Satpute et al., 2013; Craig, 2009; Damasio, 1999). Regarding the amygdala, this is the subcortical region of the brain associated with emotional intensity and the detection of a possible threat and fear (Whalen et al., 1998). It is the amygdala that translates the emotional salience in a possible threat, leaving the body on alert for flight or fight (Whalen et al., 1998). It was this mechanism in the subcortical region of amygdala that allowed organisms to survive longer in an ecosystem full of threats and predators (Whalen et al., 1998).

Emotional intelligence is the ability to think intelligently about one's own emotions, to react and act intelligently during the emotional experience, and to control and regulate emotions intelligently in a specific situation. The concept of emotional intelligence assumes that the mind can understand, reflect and regulate emotions and behaviors in an appropriate way in a given situation.

Emotional self-awareness involves several capacities: become aware of the emotion you are feeling at a given moment – anterior insula (Craig, 2009); name an emotion, in a process called emotional literacy – associated with the VLPFC (Satpute et al., 2013); and reflect on emotions, their origins in each situation and the planning of an appropriate response strategy – associated with the dorsomedial prefrontal cortex and the dorsal anterior cingulate cortex (Satpute et al., 2013).

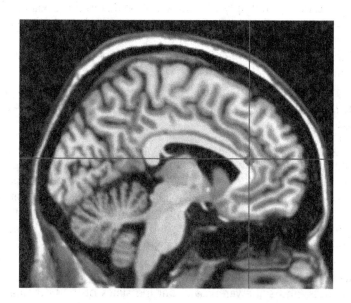

Figure 4.3 Representation of the dorsal anterior cingulate cortex associated with self-reflections about the emotional impulses, at the intersection of the horizontal line with the vertical line.

This ability to think and reflect on emotions is also a component of emotional self-awareness and contributes decisively to the development of emotional intelligence, because it is this set of reasoning and mental reflections that will allow the planning of smarter strategies in each situation.

The Role of the Dorsomedial Prefrontal Cortex on Emotional Self-Reflection

The dorsomedial prefrontal cortex is a brain region associated with evaluation and reflection on important aspects related to the self (Northoff & Bermpohl, 2004), and it is crucial for self-evaluation (Kelley et al., 2002). It has also been mentioned earlier that the dorsomedial prefrontal cortex is related to the reflections and evaluations of emotions that the person is feeling (Satpute et al., 2013).

The Importance of Evolution and Expansion of Awareness

According to one of the founders of positive psychology and one of the greatest thinkers in psychology, Professor Mihalyi Csikszentmihalyi, our society needs to help people develop in a structured and systematic way the evolving self and help people to expand consciousness (Csikszentmihalyi, 2004, 1993).

In one of his writings, entitled "Materialism and Evolution of Consciousness," he created a model based on three axioms: what we call life is a sequence of events that occur in the consciousness, that is, experiences that occur throughout

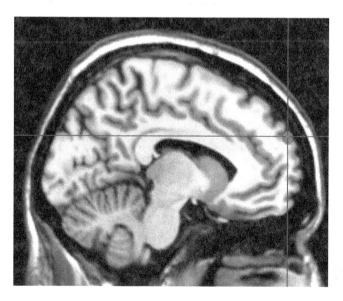

Figure 4.4 Representation of the dorsomedial prefrontal cortex associated with self-reflections about one's own emotions, at the intersection of the horizontal line with the vertical line.

a lifetime; to arise in consciousness, experiences require the consumption of psychic energy, which we call attention; and the quality and content of a person's life depend on which content people pay attention to most of their time (Csikszentmihalyi, 2004).

In fact, researchers have shown that people tend to have higher levels of well-being and happiness and lower levels of depression and anxiety when paying more attention to their conscience, in a systematic way, to intrinsic values, namely, to personal psychological growth, the establishment of meaningful and deep relationships, and significantly contributing to society, starting from the evolution of the "self for the benefit of others" (Kasser & Ryan, 1993, 1996).

On the other hand, when consciousness only devotes its attention to the extrinsic values such as social reputation, power, financial gains, fame, image and material goods, the effects are an increase in rates of depression, anxiety and emotional instability (Kasser & Ryan, 1993, 1996).

So, being emotionally intelligent also means learning to have a logic of self-evolution and expansion of consciousness, cultivating more intrinsic values compared to extrinsic values (Kasser & Ryan, 1993, 1996).

There are several ways to broaden consciousness. One is to cultivate systematically and frequently positive emotions such as joy, gratitude, admiration and moral elevation.

As mentioned earlier when discussing the broaden-and-build theory of positive emotions, positive emotions can provide resources that facilitate an expansion of consciousness, where the boundary between the self and others becomes more tenuous and the consciousness of the self expands in such a way that it includes others as part of consciousness, in the direction of a universal consciousness and to contribute significantly to the greater good (Waugh & Fredrickson, 2006).

References

Barrett, L. F., Quigley, K. S., Bliss-Moreau, E., & Aronson, K. R. (2004). Interoceptive sensitivity and self-reports of emotional experience. *Journal of Personality and Social Psychology*, *87*(5), 684. https://doi.org/10.1037/0022-3514.87.5.684

Bush, G., Luu, P., & Posner, M. I. (2000). Cognitive and emotional influences in anterior cingulate cortex. *Trends in Cognitive Sciences*, *4*(6), 215–222. https://doi.org/10.1016/S1364-6613(00)01483-2

Craig, A. D. (2002). How do you feel? Interoception: The sense of the physiological condition of the body. *Nature Reviews Neuroscience*, *3*, 655–666. https://doi.org/10.1038/nrn894

Craig, A. D. (2003). Interoception: The sense of the physiological condition of the body. *Current Opinion in Neurobiology*, *13*, 500–505. https://doi.org/10.1016/S0959-4388(03)00090-4

Craig, A. D. (2009). How do you feel-now? The anterior insula and human awareness. *Nature Reviews Neuroscience*, *10*(1), 59–70. https://doi.org/10.1038/nrn2555

Critchley, H. D., Wiens, S., Rotshtein, P., Öhman, A., & Dolan, R. J. (2004). Neural systems supporting interoceptive awareness. *Nature Neuroscience, 7*(2), 189–195. https://doi.org/10.1038/nn1176

Csikszentmihalyi, M. (1993). *The evolving self: A psychology for the third millennium.* Harpercollins.

Csikszentmihalyi, M. (2004). Materialism and the evolution of consciousness. In T. Kasser & A. D. Kanner (Eds.), *Psychology and consumer culture: The struggle for a good life in a materialistic world* (pp. 91–106). American Psychological Association.

Damasio, A. R. (1999). *Looking for spinoza: Joy, sorrow, and the feeling brain.* Harcourt.

Damasio, A. R. (2005). *Descartes' error: Emotion, reason, and the human brain.* Penguin.

Frijda, N. H., Kuipers, P., & Schure, E. (1989). Relations among emotion, appraisal, and emotional action readiness. *Journal of Personality and Social Psychology, 57*, 212–228. https://doi.org/10.1037/0022-3514.57.2.212

Kasser, T., & Ryan, R. M. (1993). A dark side of the American dream: Correlates of financial success as a central life aspiration. *Journal of Personality and Social Psychology, 65*, 410–422. https://doi.org/10.1037/0022-3514.65.2.410

Kasser, T., & Ryan, R. M. (1996). Further examining the American dream: Differential correlates of intrinsic and extrinsic goals. *Personality and Social Psychology Bulletin, 22*, 280–287. https://doi.org/10.1177/0146167296223006

Kawasaki, H., Adolphs, R., Kaufman, O., Damasio, H., Damasio, A. R., Granner, M., Bakken, H., Hori, T., & Howard, M. A. (2001). Single-neuron responses to emotional visual stimuli recorded in human ventral prefrontal cortex. *Nature Neuroscience, 4*(1), 15–16. https://doi.org/10.1038/82850

Kelley, W. M. et al. (2002). Finding the self? An event-related fMRI study. *Journal of Cognitive Neuroscience, 14*, 785–794. http://dx.doi.org/10.1162/08989290260138672

Mayer, J. D., & Salovey, P. (1997). What is emotional intelligence? In P. Salovey & D. Sluyter (Eds.), *Emotional development and emotional intelligence: Educational implications* (pp. 3–31). Basic Books.

Northoff, G., & Bermpohl, F. (2004). Cortical midline structures and the self. *Trends in Cognitive Sciences, 8*, 102–107. https://doi.org/10.1016/j.tics.2004.01.004

Satpute, A. B., Shu, J., Weber, J., Roy, M., & Ochsner, K. N. (2013). The functional neural architecture of self-reports of affective experience. *Biological Psychiatry, 73*(7), 631–638. https://doi.org/10.1016/j.biopsych.2012.10.001

Schore, A. N. (2003). *Affect Regulation and the Repair of the Self* (Norton Series on Interpersonal Neurobiology, Vol. 2). WW Norton & Company.

Waugh, C. E., & Fredrickson, B. L. (2006). Nice to know you: Positive emotions, self-other overlap, and complex understanding in the formation of new relationships. *Journal of Positive Psychology, 1*, 93–106. https://doi.org/10.1080/17439760500510569

Whalen, P. J., Bush, G., McNally, R. J., Wilhelm, S., McInerney, S. C., Jenike, M. A., & Rauch, S. L. (1998). The emotional counting stroop paradigm: A functional magnetic resonance imaging probe of the anterior cingulate affective division. *Biological Psychiatry, 44*(12), 1219–1228. https://doi.org/10.1016/S0006-3223(98)00251-0

Zaki, J., & Ochsner, K. N. (2012). The neuroscience of empathy: Progress, pitfalls and promise. *Nature Neuroscience, 15*(5), 675–680. https://doi.org/10.1038/nn.3085

Chapter 5

The Neural Bases of Emotional Self-Control

Before describing the brain mechanisms underlying emotional self-control, it is important to explain in detail what types of emotional regulation exist that were studied scientifically.

The ability to control emotional impulses is an important result from a person who is considered to have emotional intelligence. From an intrapersonal perspective, emotional self-awareness forms the basis for preparing for all other skills and capabilities of emotional intelligence, but from an intrapersonal point of view, emotional self-control is perhaps the most important competence, losing only in terms of importance to emotional self-awareness.

It is a critical competence because it is fundamental for the management of human life, and it is also crucial for the management of human relationships. Without emotional self-control, we are adrift in the complex and challenging world of modern society and human relationships.

Most of the problems we face are due to the difficulty of regulating emotions, promoting at the intrapersonal level a higher incidence of depression, anxiety, mood disorders and an increasing number of other psychopathologies, and at the interpersonal level, fostering an increase in divorces and conflicts in relationships in the most varied spheres of social life.

Helping every human being to develop effective strategies for controlling their own emotions is a necessity for all educational organizations, whether primary, secondary or higher education, and for other organizations and companies when training in personal development and in organizational skills. There are three large historical lines of scientific research that study the regulation of emotions. The pioneer is psychoanalysis, which studies the mechanisms of defense of the ego and the study of repression (more recently referred to as emotional suppression), initiated by Freud and continued by his daughter Anne Freud as well as by several other psychoanalysts (Erdelyi, 1974).

The second line of investigation is the study of coping mechanisms and stress management, which arose during the psychodynamic approach in the 1960s, in situations that exceed the personal resources of individuals. This generated the first scientific research of cognitive reappraisal, which showed that subjective responses and physiological factors tend to decrease when a movie

DOI: 10.4324/9781003503880-6

potentially generating negative emotions of a surgical procedure was analyzed by the participants analytically and objectively (Lazarus & Alfert, 1964).

The third line of research is the scientific study of self-regulation on children throughout the developmental process. This research became classically known as the delay of gratification, which is described in detail in Part II of this book about applications of emotional intelligence to the education of children and adolescents (Mischel et al., 1989). Basically, in this line of research, the ability to delay gratification is associated with several benefits for humans and constitutes a fundamental form of lifelong self-regulation and well-being in life.

Types of Emotional Regulation

Two large groups of regulatory strategies were fundamentally identified: one behavioral and the other cognitive (Gross, 2002). In the behavioral strategy of emotion regulation, the goal is to suppress the expression of the emotional response, also called suppressive emotional regulation (Jackson et al., 2000).

The limitation of this strategy is that despite repressing an unsuitable emotional response, it does not eliminate the negativity of the emotional experience, decreasing the ability to commit to memory and increasing the activation of the sympathetic nervous system (Gross, 2002; Jackson et al., 2000).

The other strategy is cognitive, one that generates an emotional response or one that modifies an already activated emotional response (Ochsner & Gross, 2005).

Behavior Strategy for Emotional Regulation: Suppression

Suppressive emotional regulation involves attempting to ignore, avoid and hide negative emotions because they are assessed as threatening. It is what Freud called "repression" (Freud, 1966). Suppressive emotional regulation occurs at the beginning of a sequence when people deny or ignore the experience or the existence of emotional stimulation. Thus, the deletion includes avoiding the emotional experience or, in a more subtle modality, distancing to minimize the impact of the emotional experience.

Consequently, the emotion felt is not fully brought to consciousness, for there is little inner exploration of the emotional experience. Some studies have recorded that emotional suppression is related to higher levels of depression (Berenbaum et al., 1999). Furthermore, as the emotional experience is not entirely brought to consciousness, it tends, later, to rise from the unconscious to cause the phenomenon of emotional rumination (Thomsen et al., 2011).

Because of its negative consequences from an intrapersonal point of view, the suppression cannot be considered an emotionally intelligent strategy of management and regulation of emotions. That is why it is not considered here in as much detail as other emotional regulation strategies.

In a study of the neural bases of suppression, the researchers identified two aspects of the mental process of emotional suppression: a pathway of so-called inhibitory control composed of the regions of the dorsolateral prefrontal cortex, which regulates representations of the sensory effects of memory (i.e., the thalamus), and the regions of the visual cortex, fusiform gyrus and pulvinar nucleus of the thalamus; the other pathway, the prefrontal cortex, regulates the regions of the hippocampus and amygdala. Both are influenced by the anterior prefrontal cortex (Depue et al., 2007).

Cognitive Strategy of Emotion Regulation: Cognitive Reappraisal

The cognitive regulation strategies can be used to generate emotional responses or to regulate a negative emotional response already activated (Ochsner & Gross, 2005). The use of this strategy has been studied based on the following three approaches.

The first approach has been to identify the neural bases of anticipatory responses preceding the expected emotional events. This anticipation is associated in several neuroscientific studies with the dorsomedial prefrontal cortex (Hsieh et al., 1999; Ploghaus et al., 1999; Porro et al., 2002).

In addition to this region, the anticipation of emotional events tends to activate the dorsal anterior cingulate cortex, which, according to what has already been mentioned, is responsible for the expression of anxiety (Critchley et al., 2004) by assessing the meaning of an imminent threat (Etkin et al., 2011) and the development and monitoring of an emotional or cognitive conflict (Etkin et al., 2006).

As mentioned earlier, the dorsomedial prefrontal cortex and the dorsal anterior cingulate cortex are responsible for the reflection about emotions (Satpute et al., 2013). It is understandable that during the anticipation of an emotionally salient event, there is a reflection on emotions and what the individual expects, which causes these regions to be activated.

The second approach studies how expectations about stimuli influence emotional responses and their respective neural bases (Ochsner & Gross, 2005). For example, one study showed that a non-pain stimulus was perceived as painful when individuals had the expectation that they would feel pain, and this activated the anterior cingulate cortex and the posterior insula, probably associated with emotions connected with the expectation of pain (Sawamoto et al., 2000). In another study, the same non-pain stimulus activated the ventral anterior cingulate cortex (vACC)/rostral and the medial temporal lobe, probably associated with cognitive expectation of pain (Ploghaus et al., 2001).

The third approach contrasts the origin of top-down responses generated by mental beliefs about a stimulus with bottom-up responses generated by the direct perception of a stimulus, without mental mediation (Ochsner & Gross,

2005). So far, only one study has carried out this contrast between top-down and bottom-up responses, and it turned out that although in both responses the amygdala has been activated, only the top-down response activated the anterior cingulate cortex, lateral prefrontal cortex and MPFC, i.e., it was a cognitive response that generated an aversive evaluation (unpleasant) in the face of a neutral stimulus (Ochsner & Gross, 2004). In this study, participants were asked to see aversive images (bottom-up) or to think of neutral images in a negative way (Ochsner & Gross, 2004). It has been shown that even in the face of neutral stimuli, our mind can create an aversive response and activate the same regions when the stimulus is truly aversive and unpleasant (Ochsner & Gross, 2004).

Cognitive Reappraisal

The main strategy of emotional regulation is cognitive reappraisal (Ochsner & Gross, 2005), which is the ability to reinterpret the meaning of a stimulus with the objective of modifying the emotional response to this stimulus, in such a way that the response is not so salient or negative (Gross, 1998).

Currently, there is a lot of scientific evidence of the success of emotional, cognitive and behavioral aspects of cognitive reappraisal when compared to the suppressive regulation of emotions (Gross, 2015). The advantage of cognitive reappraisal is the moment in which it is used, i.e., the fact that it starts before the emotional response.

THE ROLE OF THE DORSOLATERAL PREFRONTAL CORTEX IN COGNITIVE REAPPRAISAL

One study demonstrated that cognitive reappraisal activates the dorsolateral prefrontal cortex and the medial orbitofrontal cortex (Ochsner et al., 2002).

The activation of the orbitofrontal cortex region is associated with the representation of the positive or negative emotional effects of a stimulus (O'Doherty et al., 2001) and is modulated by cognitive reappraisal (Ochsner et al., 2002). This means that the cognitive reappraisal (associated with the dorsolateral prefrontal cortex) may modify the emotional representation of a negative stimulus for a positive representation, which is associated with the medial orbitofrontal cortex (Ochsner et al., 2002).

One study found that cognitive reappraisal, to regulate a negative emotional response, activated the dorsal anterior cingulate cortex, the dorsomedial prefrontal cortex, the dorsolateral prefrontal cortex and the VLPFC, and it deactivated the regions of the limbic system, in particular, the NAcc and the amygdala (Phan et al., 2005). As mentioned earlier, the dorsal anterior cingulate cortex and the dorsomedial prefrontal cortex are two regions responsible for reflection about emotional stimulus (Satpute et al., 2013), and the dorsolateral prefrontal cortex is at the origin of the cognitive control (Ochsner et al., 2002).

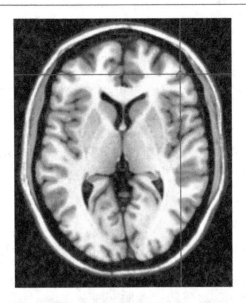

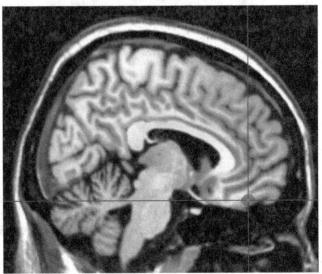

Figure 5.1 Representation of (a) the dorsolateral prefrontal cortex and (b) the medial orbitofrontal cortex associated with emotional self-control, at the intersection of the horizontal line with the vertical line

These data suggest that the regions of the prefrontal cortex (dorsomedial and orbitofrontal lateral) exercised control of the deactivation of the subcortical regions, namely, the NAcc and the amygdala. This is the neural mechanism of cognitive control over a negative emotional response (Phan et al., 2005).

In another important study of the neural bases of emotional self-control, researchers quantified the answers to two key questions when presenting the participants with a series of healthy foods and appetizing foods (Hare et al., 2009): (1) how much the participants, on a scale from 1 to 5, would like to eat this food, regardless of whether it is healthy or not (measure of subjective value of the food); and (2) how much the participants, on a scale from 1 to 5, choose to eat healthy foods and reject unhealthy foods (self-control measure).

These two questions were asked after the presentation of each food, whether healthy or appetizing but not healthy (Hare et al., 2009). The results were interesting. The more participants evaluated with a high score the answer to the first question (how much they would like to eat this food), i.e., attributed a higher subjective value to the food, the region of the ventromedial prefrontal cortex (vmPFC) was activated (Hare et al., 2009). These data suggest that the subjective value that is attributed to something relates to the vmPFC (Hare et al., 2009).

When the researchers asked the second question – how much the participants, on a scale from 1 to 5, choose to eat healthy foods and reject unhealthy foods (self-control measure) for each food presented – in the participants who rated high values, the dorsolateral prefrontal cortex was activated, as this was a measure of self-control (Hare et al., 2009).

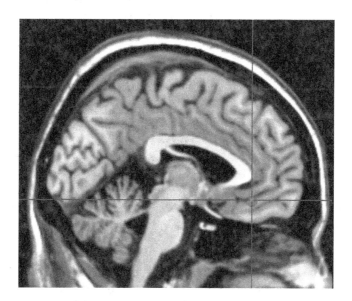

Figure 5.2 Representation of the ventromedial prefrontal cortex associated with the subjective value of an object, at the intersection of the horizontal line with the vertical line.

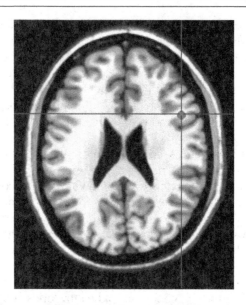

Figure 5.3 Representation of the dorsolateral prefrontal cortex, associated with self-control, at the intersection of the horizontal line with the vertical line.

The Role of the Ventral Anterior Cingulate Cortex in Emotional Self-Control The vACC is composed of the subgenual anterior cingulate cortex and pregenual anterior cingulate cortex (pACC). These regions have strong structural and functional connectivity with the anterior insula, the amygdala, the periaqueductal gray matter, the NAcc and the hypothalamus and areas of the limbic system, and they have exits to the autonomous, visceromotor and endocrine systems.

The main functions of the vACC are evaluation of the salience (i.e., importance) of emotional and motivational information and regulation of emotional responses (Vogt et al., 1992; Devinsky et al., 1995; Drevets & Raichle, 1998; Whalen et al., 1998; Ochsner & Gross, 2005; Etkin et al., 2006). A group of researchers demonstrated that the resolution of an emotional conflict is associated with activation of the pACC (rostral), and activation of this area tends to regulate the decrease in the activation of the amygdala, which usually is activated in the face of an emotional conflict (Etkin et al., 2006).

Another study showed that in the face of a stimulus from a distant threat, activation of the vACC and MPFC occurs. The researchers speculated that since the threat was distant, this allowed mental planning of adaptive responses to the situation, suggesting that the intervention of the vACC was the regulation of negative emotions (Mobbs et al., 2009).

In the classical studies of fear conditioning, it was found that the extinction of the conditioned fear is associated with activation of the vACC, and that the activation of this region modulated the decrease of fear activation in the amygdala, which presupposes a role of regulator of negative emotions (Milad et al., 2008; Phelps et al., 2004).

The human brain protects the processing of stimuli relevant to a task from mental (which researchers designate cognitive conflict) or emotional interference (i.e., emotional conflict) through mechanisms of skew of attention, to allow human beings to perform the task efficiently.

One study found that when an emotional interference (i.e., emotional conflict) appears in the performance of a task, the vACC is the regulator of this emotional interference for an efficient focus on the task, particularly the pregenual or rostral region (Egner et al., 2008). This research also suggests the role of the vACC in regulating negative emotional interferences (Egner et al., 2008).

The vACC and the vmPFC are also involved in two strategies of emotional regulation: emotional literacy, that is, the ability to name emotions in relation to facial expressions of other people, allowing a decrease in the activation of the amygdala (Lieberman et al., 2007; Satpute et al., 2013), or when the person uses a strategy of personal distraction in the face of a conditioned fear stimulus (Delgado et al., 2008).

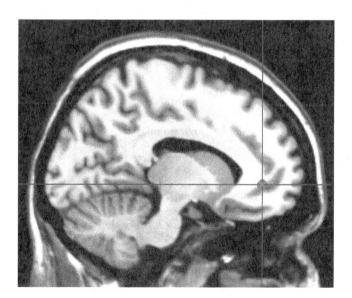

Figure 5.4 Representation of the ventral anterior cingulate cortex, associated with self-control of emotional and cognitive interference, at the intersection of the horizontal line with the vertical line.

These data suggest that the vACC and the vmPFC are regions responsible for inhibiting the emotional activation of the amygdala and the negative emotional interference during a task (Schiller & Delgado, 2010).

References

Berenbaum, H., Raghavan, C., Le, H. N., Vernon, L., & Gomez, J. (1999). Disturbances in emotion. In D. Kahneman, E. Diener, & N. Schwarz (Eds.), *Well-being: The foundations of hedonic psychology* (pp. 267–287). Sage.

Critchley, H. D., Wiens, S., Rotshtein, P., Öhman, A., & Dolan, R. J. (2004). Neural systems supporting interoceptive awareness. *Nature Neuroscience, 7*(2), 189–195. https://doi.org/10.1038/nn1176

Delgado, M. R., Nearing, K. I., LeDoux, J. E., & Phelps, E. A. (2008). Neural circuitry underlying the regulation of conditioned fear and its relation to extinction. *Neuron, 59*(5), 829–838. https://doi.org/10.1016/j.neuron.2008.06.029

Depue, B. E., Curran, T., & Banich, M. T. (2007). Prefrontal regions orchestrate suppression of emotional memories via a two-phase process. *Science, 317*(5835), 215–219. https://doi.org/10.1126/science.1139560

Devinsky, O., Morrell, M. J., & Vogt, B. A. (1995). Contributions of anterior cingulate cortex to behaviour. *Brain, 118*(1), 279–306. https://doi.org/10.1093/brain/118.1.279

Drevets, W. C., & Raichle, M. E. (1998). Reciprocal suppression of regional cerebral blood flow during emotional versus higher cognitive processes: Implications for interactions between emotion and cognition. *Cognition and Emotion, 12*, 353–385. https://doi.org/10.1080/026999398379646

Egner, T., Etkin, A., Gale, S., & Hirsch, J. (2008). Dissociable neural systems resolve conflict from emotional versus non emotional distracters. *Cerebral Cortex, 18*(6), 1475–1484. https://doi.org/10.1093/cercor/bhm179

Erdelyi, M. H. (1974). A new look at the New Look: Perceptual defense and vigilance. *Psychological Review, 81*(1), 1–25. https://doi.org/10.1037/h0035852

Etkin, A., Egner, T., & Kalisch, R. (2011). Emotional processing in anterior cingulate and medial prefrontal cortex. *Trends in Cognitive Sciences, 15*(2), 85–93. https://doi.org/10.1016/j.tics.2010.11.004

Etkin, A., Egner, T., Peraza, D. M., Kandel, E. R., & Hirsch, J. (2006). Resolving emotional conflict: A role for the rostral anterior cingulate cortex in modulating activity in the amygdala. *Neuron, 51*(6), 871–882. https://doi.org/10.1016/j.neuron.2006.07.029

Freud, S. (1966). Pre-psycho-analytic publications and unpublished drafts. In J. Strachey (Ed.), *The standard edition of the complete psychological works of sigmund freud* (vol. 1, pp. 117–128). Hogarth.

Gross, J. J. (1998). Antecedent-and response-focused emotion regulation: Divergent consequences for experience, expression, and physiology. *Journal of Personality and Social Psychology, 74*(1), 224. https://doi.org/10.1037/0022-3514.74.1.224

Gross, J. J. (2002). Emotion regulation: Affective, cognitive, and social consequences. *Psychophysiology, 39*(3), 281–291. https://doi.org/10.1017/S0048577201393198

Gross, J. J. (2015). Emotion regulation: Current status and future prospects. *Psychological Inquiry, 26*, 1–16. https://doi.org/10.1080/1047840X.2014.940781

Hare, T. A., Camerer, C. F., & Rangel, A. (2009). Self-control in decision-making involves modulation of the vmPFC valuation system. *Science, 324*(5927), 646–648. https://doi.org/10.1126/science.1168450

Hsieh, J. C., Meyerson, B. A., & Ingvara, M. (1999). PET study on central processing of pain in trigeminal neuropathy. *European Journal of Pain, 3*(1), 51–65. https://doi.org/10.1016/S1090-3801(99)90188-X

Jackson, D. C., Malmstadt, J. R., Larson, C. L., & Davidson, R. J. (2000). Suppression and enhancement of emotional responses to unpleasant pictures. *Psychophysiology, 37*(4), 515–522. https://doi.org/10.1111/1469-8986.3740515

Lazarus, R. S., & Alfert, E. (1964). Short-circuiting of threat by experimentally altering cognitive appraisal. *The Journal of Abnormal and Social Psychology, 69*(2), 195. https://doi.org/10.1037/h0044635

Lieberman, M. D., Eisenberger, N. I., Crockett, M. J., Tom, S. M., Pfeifer, J. H., & Way, B. M. (2007). Putting feelings into words. *Psychological Science, 18*(5), 421–428. https://doi.org/10.1111/j.1467-9280.2007.01916.x

Milad, M. R., Orr, S. P., Lasko, N. B., Chang, Y., Rauch, S. L., & Pitman, R. K. (2008). Presence and acquired origin of reduced recall for fear extinction in PTSD: Results of a twin study. *Journal of Psychiatric Research, 42*(7), 515–520. https://doi.org/10.1016/j.jpsychires.2008.01.017

Mischel, W., Shoda, Y., & Rodriguez, M. I. (1989). Delay of gratification in children. *Science, 244*(4907), 933–938. https://doi.org/10.1126/science.2658056

Mobbs, D., Marchant, J. L., Hassabis, D., Seymour, B., Tan, G., Gray, M., Petrovic, P., Dolan, R. J., & Frith, C. D. (2009). & Frith, C. D. (2009). From threat to fear: The neural organization of defensive fear systems in humans. *Journal of Neuroscience, 29*(39), 12236–12243. https://doi.org/10.1523/JNEUROSCI.2378-09.2009

O'Doherty, J. O., Kringelbach, M. L., Rolls, E. T., Hornak, J., & Andrews, C. (2001). Abstract reward and punishment representations in the human orbitofrontal cortex. *Nature Neuroscience, 4*, 95–102. https://doi.org/10.1038/82959

Ochsner, K. N., Bunge, S. A., Gross, J. J., & Gabrieli, J. D. (2002). Rethinking feelings: An FMRI study of the cognitive regulation of emotion. *Journal of Cognitive Neuroscience, 14*(8), 1215–1229. https://doi.org/10.1162/089892902760807212

Ochsner, K. N., & Gross, J. J. (2004). Thinking makes it so: A social cognitive neuroscience approach to emotion regulation. *Handbook of Self-Regulation: Research, Theory, and Applications,* 229–255.

Ochsner, K. N., & Gross, J. J. (2005). The cognitive control of emotion. *Trends in Cognitive Sciences, 9*(5), 242–249. https://doi.org/10.1016/j.tics.2005.03.010

Phan, K. L., Fitzgerald, D. A., Nathan, P. J., Moore, G. J., Uhde, T. W., & Tancer, M. E. (2005). Neural substrates for voluntary suppression of negative affect: A functional magnetic resonance imaging study. *Biological Psychiatry, 57*(3), 210–219. https://doi.org/10.1016/j.biopsych.2004.10.030

Phelps, E. A., Delgado, M. R., Nearing, K. I., & LeDoux, J. E. (2004). Extinction learning in humans: Role of the amygdala and vmPFC. *Neuron, 43*(6), 897–905. https://doi.org/10.1016/j.neuron.2004.08.042

Ploghaus, A., Narain, C., Beckmann, C. F., Clare, S., Bantick, S., Wise, R., Matthews, P. M., Nicholas, J., Rawlins, P., & Tracey, I. (2001). Exacerbation of pain by anxiety is associated with activity in a hippocampal network. *Journal of Neuroscience, 21*(24), 9896–9903. https://doi.org/10.1523/JNEUROSCI.21-24-09896.2001

Ploghaus, A., Tracey, I., Gati, J. S., Clare, S., Menon, R. S., Matthews, P. M., & Rawlins, J. N. P. (1999). Dissociating pain from its anticipation in the human brain. *Science, 284*(5422), 1979–1981. https://doi.org/10.1126/science.284.5422.1979

Porro, C. A., Baraldi, P., Pagnoni, G., Serafini, M., Facchin, P., Maieron, M., & Nichelli, P. (2002). Does anticipation of pain affect cortical nociceptive systems? *Journal of Neuroscience, 22*(8), 3206–3214. https://doi.org/10.1523/JNEUROSCI.22-08-03206.2002

Satpute, A. B., Shu, J., Weber, J., Roy, M., & Ochsner, K. N. (2013). The functional neural architecture of self-reports of affective experience. *Biological Psychiatry, 73*(7), 631–638. https://doi.org/10.1016/j.biopsych.2012.10.001

Sawamoto, N., Honda, M., Okada, T., Hanakawa, T., Kanda, M., Fukuyama, H., Konishi, J., & Shibasaki, H. (2000). Expectation of pain enhances responses to nonpainful somatosensory stimulation in the anterior cingulate cortex and parietal operculum/posterior insula: An event-related functional magnetic resonance imaging study. *Journal of Neuroscience, 20*(19), 7438–7445. https://doi.org/10.1523/JNEUROSCI.20-19-07438.2000

Schiller, D., & Delgado, M. R. (2010). Overlapping neural systems mediating extinction, reversal, and regulation of fear. *Trends in Cognitive Sciences, 14*(6), 268–276. https://doi.org/10.1016/j.tics.2010.04.002

Thomsen, D. K., Tønnesvang, J., Schnieber, A., & Olesen, M. H. (2011). Do people ruminate because they haven't digested their goals? The relations of rumination and reflection to goal internalization and ambivalence. *Motivation and Emotion, 35*, 105–117. https://doi.org/10.1007/s11031-011-9209-x

Vogt, B. A., Finch, D. M., & Olson, C. R. (1992). Functional heterogeneity in cingulate cortex: The anterior executive and posterior evaluative regions. *Cerebral Cortex, 2*(6), 435–443. https://doi.org/10.1093/cercor/2.6.435-a

Whalen, P. J., Bush, G., McNally, R. J., Wilhelm, S., McInerney, S. C., Jenike, M. A., & Rauch, S. L. (1998). The emotional counting stroop paradigm: A functional magnetic resonance imaging probe of the anterior cingulate affective division. *Biological Psychiatry, 44*(12), 1219–1228. https://doi.org/10.1016/S0006-3223(98)00251-0

Chapter 6

The Science of Human Relationships and Its Relationship With Neuroscience

Every human relationship must be analyzed as a single system. The relationship of a father with a child constitutes a system, and the relationship of that same father with his wife forms another completely distinct system. Therefore, it is important to analyze the elements that constitute patterns of behaviors, thoughts and emotions that influence relationships in a healthy way and to which absence can generate many conflicts, difficulties and dysfunctions in relationships.

Relational Systems

It is proposed in this book that there exist two possible systems of relations with others: a symbiotic relationship system and another relationship that promotes psychological maturity.

The first is dysfunctional, and the second is functional and tends to promote the evolution of the self of both members of the relationship.

Symbiotic Relationship System

All kinds of relationships create dependency and do not consider the development of the other or the development of the real self of each of the relationship members (Bauman, 2001).

When you establish a relationship based on fantasies, people do not tend to develop love for the other but to seek only to satisfy their personal illusions. Therefore, they are more self-centered when in the relationship (Bauman, 2001). The result is simple: the more focused a person is in their own personal fantasies and illusions (false needs), the more symbiotic the relationship will tend to be, and the less psychological growth and long-term happiness will be obtained by both members.

Symbiotic relationships are based on a selfish perspective of relationships because people do not think of both parts but only the satisfaction of their own illusions (Bauman, 2001). And when people satisfy their fantasies, they tend to create an illusion that they "love" that person. This is pure illusion.

This occurs in love relationships between friends, parents, and children, and in other types of relationships.

This erroneous assessment occurs because people do not reflect deeply about themselves, with a high level of emotional self-awareness and evolution of the self, nor do they seek to know the other in a profound and real way. This represents a lack of depth of the self and, consequently, leads to a superficial analysis of human relationships (Bauman, 2003, 2001). Thus, while the illusion endures, people feel "happy" and "full of love."

The risk that people have in relating to someone they do not seek to know in a profound way is to learn late that this person does not bring something good for you. Here comes disappointment. But it is only disappointing for the person who has been deluded before. The search to know oneself proves to be as important as knowing others as they are. It is simple.

However, it seems so complicated in real life because people tend to prefer to live immersed in fantasies that our culture has instilled in us since childhood than to seek to know one's own real self and to know the real self of others (Bauman, 2001, 2003). This is an insurance policy for any wedding.

Of course, people often never show who they really are. But it is up to you, to us, in any relationship, to learn to know others in more depth. And the first step is to eliminate our own fantasies and illusions (Bauman, 2001).

This is for any human relationship. With our parents, brothers, family members, friends, loving partners, children, with anyone. When it comes to any relationship, all human beings must first ask themselves: Why do I relate to this person? What motivates me in this relationship? Will it contribute to my development and the development of the other? Will it just be a superficial exchange of sensations and Shakespearean affective fantasies or the deeper sharing of my real self with the other real self? What is the person really looking for?

If humans seek relationships based on illusory motivations and values, certainly the consequence will be an enormous frequency of illusions, disappointments and conflicts, and the relationships will be marked by a lot of emotional instability (Bauman, 2003). These certainly are not emotionally intelligent decisions. The experience of fantasy, from an emotional point of view, tends to generate very intense emotions of pleasure (pride and vanity) mixed with euphoria and mania; in the end, fantasy tends to generate sadness, anger and depression.

Everything tends to be a big illusion. When seeking a healthy process of psychological maturation, human beings will conclude that they do not need the illusions to be authentically happy, because they will base happiness on their real self and the real self of others.

When they experience illusory values, symbiotic relationships tend to have two possible positions: a dominator and a dominated one. The first is the one who, in the relational system, is the bearer of your partner's fantasy; the second will be the one who will have more illusions, whether fantasies of the ego or illusions of superiority or inferiority. The symbiotic dysfunctional relationship is only broken if the dominated partner awakens from the illusions and frees themself from the toxic relationship, or if, by some miracle, the dominant partner finds maturity and frees themself from the illusion of superiority and the desire to dominate their partner to satisfy their fantasies.

To lead a symbiotic relationship toward the psychological growth of both parts, it is essential that the starting point is to bring both to reality.

From there, the two parts will be able to identify whether the relationship has any positive functional potential in the future. After the "bath" of reality, the self-knowledge and deep knowledge of the other become indispensable in assessing the real potential of the relationship and whether the real motivations remain. In most cases, the illusion is so great that the other partner might get angry at themself for "not having seen the reality before." Or the partner will leave the relationship, tending to victimize themself, and become "depressed" when they realize that the responsibility of the choice was theirs. Usually, the partner will tend to blame the other for their disillusionment and victimize themself, when, in fact, they cultivated the illusion themself.

Psychological Maturity System

The main objective of a relationship of psychological maturity is the development of the real self of both parts, toward evolution of self and complexity. A relationship of psychological maturity implies a genuine concern for the other, in helping the other in their true needs and contributing to the other developing their self, in a logical reciprocity, not only of material resources but, above all, of moral resources, of values, of purposes of life and of genuine and true positive feelings.

The exchange of all these psychological resources is very important in encouraging the self to develop. And it is very gratifying to experience relationships with these characteristics, whether a romantic relationship, a friendship relationship, a relationship between parents and children or a relationship between teacher and student. The following are four fundamental conditions necessary for a mature relationship to occur, leading to the complexity of and evolution of both parts:

Care, attention, love, and genuine concern for the well-being of the other – It is often said that we only truly know someone at the moments of greater difficulty. That is, when we find ourselves in more difficulty, then it becomes clear who genuinely cares about us and really wishes us well. It is something we perceive emotionally. It is something that we feel. The genuineness of someone's concern for us is something you feel. And feeling that someone really cares about us is very rewarding. It generates gratitude, affection and love.

Genuine love and concern for someone is one of the best things we can cultivate in a relationship. And this behavioral, mental attitude and emotional concern toward each other can potentially generate a positive spiral relationship, which allows them to be, on the one hand, extremely useful for their own personal evolution; on the other hand, it can generate frequent positive emotions, well-being and happiness between both parts.

Erich Fromm, in his masterpiece *The Art of Loving*, describes the question as follows:

The correct quotation is this:

> Most people see the problem of love primarily as that of being loved, rather than that of loving, of one's capacity to love. Hence the problem to them is how to be loved, how to be lovable.
>
> (Fromm, 1956, pp. 1–2)

And Fromm continues:

> Many of the ways to make oneself lovable are the same as those used to make oneself successful to "win friends and influence people." As a matter of fact, what most people in our culture mean is essentially a mixture between being popular and having sex appeal.
>
> (Fromm, 1956, pp. 1–2)

> Love is an activity, not a passive affect; it is a "standing in," not a "falling for." In the most general way, the active character of love can be described by stating that love is primarily giving, not receiving.
>
> (Fromm, 1956, p. 22)

And he continues:

> For the productive character, giving has an entirely different meaning. Giving is the highest expression of potency. In the very act of giving, I experience my strength, my wealth, power. This experience of heightened vitality and potency fills me with joy. I experience myself as overflowing, spending, alive, hence as joyous. Giving is more joyous than receiving, not because it is deprivation, but because the act of giving lies the expression of my aliveness.
>
> (Fromm, 1956, p. 23)

And later he continues:

> The essence of love is to "labor" for something and "to make something grow," that love and labor are inseparable. One loves that for which one labors, and one labors for that which one loves.
>
> (Froom, 1956, p. 27)

Deep knowledge of the other – One of the biggest problems of relationships is the superficiality in seeking to know the other as they are, the same way one should seek to know oneself. The higher knowledge of the other, and this knowledge leads to an adjustment toward the progress of each person's self, the relationship, in addition to being more emotionally satisfying, produces complexity and depth, and this will be progressively and more intrinsically gratifying. We cannot create a relationship of mutual growth if there is no concern in knowing each other in depth. One of the criticisms of modern societies is superficiality and selfishness (due to illusory values) in relationships, and as a result, these characteristics produce great suffering (Bauman, 2003).

Responsibility in the relationship with the other – Erich Fromm says the following about responsibility:

> Today, responsibility is often meant to denote duty, something imposed upon from the outside. But responsibility, in its true sense, is an entirely voluntary act; it is my response to the needs, expressed or unexpressed, of another human being. To be responsible is to be ready to respond.
> (Fromm, 1956, pp. 27–28)

Therefore, responsibility is the ability to respond to the fundamental needs of the other: competence, affiliation, benevolence and autonomy and any material needs.

Respect – Erich Fromm writes about respect:

> Respect means the concern that the other person should grow and unfold as he is. Respect, thus, implies the absence of exploitation. I want the loved person to grow and unfold for his own sake, and in his own ways, and not for the purpose of serving me. If I love the person, I feel one with him or her, but with him as he is, not as I need him to be as an object for my use.
> (Fromm, 1956, p. 28)

What is the case in "modern relationships" is a huge and widespread lack of respect. We are not referring to "social respect" as British politeness, which refers to a conventional morality of being educated, but respect by the human being, by the characteristics of their real self, of not trampling a person's individuality and helping the other to develop based on their real self, according to their unique characteristics, their own pace of development and their own choices, even if we do not agree with them. Respecting the other is not imposing what we would like the other to be but is respecting the other as they are. There is a common tendency, when people reach a greater degree of intimacy, to want to violate the space of the other and impose one's vision on the other, without having respect for the other's real self. At present, the lack of respect is a frequent source of many relational conflicts.

The word *respect* means *respiscere*, that is, "look at," which means dealing with the other based on their real self, and not the self that we would like the other to be. Relationships exist for human beings to learn to evolve together, to share the evolution of each real self, to learn from each other, expanding and exchanging mental, emotional, moral and material resources.

Nowadays, there is a very selfish view of relationships, and these selfish values see relationships only as sources of satisfaction of our own fantasies and illusory desires, which tends to cause numerous relational problems and, consequently, emotional instability, negative emotions and depression (Bauman, 2001). The establishment of relationships based on illusory motivations and values is not an emotionally intelligent choice. These latest statements are confirmed by the substantial increase in the divorce rate (Stevenson & Wolfers, 2007), decline in marriage satisfaction in the last 30 years (Myers, 2001), increase in the rate of loneliness (Cacioppo & Patrick, 2008) and decrease in the number of close friends, on average, from three to two in the United States between 1985 and 2004 (McPherson et al., 2006), as well as the fact that American children spend, on average, more time watching TV than talking to their parents (Vandewater et al., 2006).

These data suggest the importance of intrinsic values in relationships to help human beings not be bound by the illusions of materialism and illusory relationships (Ryan et al., 1994; Blais et al., 1990) and suffer the consequences of emotional instabilities like anger, sadness, anxiety and depression (Kasser, 2016) and learn how to cultivate healthy relationships that are conducive to mutual progress and happiness, without childish illusions or fantasies.

Neuroscience, Biology and Human Relationships

From the point of view of evolutionary psychology, ethology and evolutionary anthropology, researchers unanimously claim that human beings and their ancestors are a highly biologically prepared species for social life. Social life and prosocial emotions are fundamental needs of our species as well as of our primate ancestors (de Waal, 2016).

The famous primatologist Frans de Wall, throughout his career, demonstrated the origins of empathy, morality and altruism in our primate ancestors (de Waal, 2016). Humans have a biological structure that facilitates an intense social life.

Our biological structure of mammals makes our children, from a biological point of view, extremely dependent on mothers during the first years of life, more than most species, in order to mature the whole, more complex, as compared to other species, biological apparatus.

Thus, the biological mother's monitoring of her child requires a whole structure that facilitates affiliation and the strong social and emotional connection with the child.

The Relation of Oxytocin With Emotions and Social Relationships

Oxytocin is a hormone produced by the hypothalamus and stored in the pituitary gland. It has been known for a long time for its role in facilitating muscle contractions and reducing bleeding during childbirth as well as in the stimulation of breast milk production (Lee et al., 2009).

A group of neuroeconomists and psychologists from the University of Zurich did an experiment with two groups, one control group and the other group that received a sufficient amount of oxytocin inserted intranasally. They both held a game of trust in which money was involved. The results showed that the group of participants who received a dose of oxytocin tended to trust more the other participants in comparison with the control group. This study, published in the scientific journal *Nature*, was able to demonstrate a cause-and-effect relationship of oxytocin on people's trust (Kosfeld et al., 2005). This paper became so famous that the media hastened to call oxytocin the love hormone. In addition, another study revealed that oxytocin tends to reduce anxiety and social stress (Heinrichs et al., 2003).

This "miraculous" idea of oxytocin as the love hormone was further reinforced by another study in which the researchers demonstrated that the intake of oxytocin increased people's cognitive empathy, the ability to infer the mental state of others based on complex social cues (Domes et al., 2007). Another study showed that oxytocin tends to increase empathy dependent on the activation of the amygdala (Hurlemann et al., 2010), to facilitate contact between people during social interactions (Guastella et al., 2008) and to improve coordination and cooperation between two people (Arueti et al., 2013).

In addition, heterosexual participants with a stable loving relationship and who received a dose of oxytocin tended to remain at a significant distance of sexually attractive women compared to the group control (composed of heterosexual participants with a loving relationship and who did not receive oxytocin). In addition, these latter participants obtained a "less reflective" response when exposed to erotic images of a beautiful woman, through an experimental paradigm specific to that effect (Scheele et al., 2012). This latest study suggests that oxytocin may increase the monogamous fidelity behavior between loving partners.

Another study demonstrated that a social interaction in which there is harmony between people tends to increase the internal levels of oxytocin on both sides and that intranasal intake of oxytocin can increase the expressiveness of both happiness and fear in facial and vocal expressions (Spengler et al., 2017). A newer study has shown that nasal intake of oxytocin increased the prosocial behavior of donation of individualistic participants, modulated by the representation of the social value of the donation in the amygdala, compared to participants previously considered altruistic (Liu et al., 2019). This study suggests that oxytocin promotes altruistic behavior, especially in people more selfish, and that the difference is that the amygdala is more activated in this case (Liu et al., 2019).

All this excitement surrounding the "miracle of oxytocin" turned out to be a little precipitated, because some time later, studies emerged that demonstrated that oxytocin does not always promote prosocial behaviors and feelings. Researchers identified that in some situations, oxytocin may even amplify a series of aversive behaviors. For example, oxytocin tends to increase social anxiety in both humans (Eckstein et al., 2014) and animals (Duque-Wilckens et al., 2018).

Another very interesting study showed that oxytocin tends to increase endogroup favoritism and exogroup prejudice and harm through increased financial aid to those who belong to the same group, and greater financial loss for those who do not belong to the reference group (De Dreu et al., 2010). In addition, in a new study led by the same investigator, it has been shown that oxytocin may increase ethnocentrism, i.e., the tendency to see the group itself as more important and superior to other groups (De Dreu et al., 2011). Based on this latest research, the idea that oxytocin is a hormone that produces prosocial relationships toward others began to be questioned (Van Anders et al., 2013).

To try to clarify the two contradictory groups of results, some researchers have proposed an alternative explanation for the effects of oxytocin, in which the latter tends to increase the salience of social contexts, whether aversive or prosocial (Shamay-Tsoory & Abu-Akel, 2016). According to these researchers, in addition to increasing the salience of social contexts, oxytocin is highly connected with the dopaminergic process (i.e., dopamine release) of the brain's reward system, namely, the septal area, the NAcc, the ventral tegmental area (VTA) and the pallidum. They also state that there is a huge individual variability in the effects of oxytocin to produce prosocial or aversive behaviors, dependent on temperament, personality, gender or possible degree of psychopathology (Shamay-Tsoory & Abu-Akel, 2016; Bartz et al., 2011).

The VTA, the septal area, NAcc and the pallidum are associated with prosocial moral motivation within the brain reward system (Zahn et al., 2020). The interaction between all these factors will help predict whether the individual under the effect of increased oxytocin will develop prosocial or aversive behaviors, although the social context is also an important explanatory factor in this process (Shamay-Tsoory & Abu-Akel, 2016; Bartz et al., 2011).

When the approach and avoidance motivations were mentioned as the two motivational mechanisms most suitable to explain the asymmetry between the left and right prefrontal cortex, compared to the previous idea of positive emotions associated with the left hemisphere and negative emotions associated with the right, the idea here is similar. Oxytocin promotes both the approach and aversive motivations, depending on the social context, personality, temperament and intra-individual differences between people

(Shamay-Tsoory & Abu-Akel, 2016; Bartz et al., 2011). These data are important so that you do not fall into the euphoria of thinking that oxytocin is a love hormone, and that, in future studies, caution is used so people do not jump to conclusions. So far, the data suggest that oxytocin only amplifies some predisposition existing in the real self of individuals and social contexts.

The Meaning of Love in Relationships

If we assume that the concept of love is just an emotion characterized by tenderness toward the other, or by the "enchantment" of being "in love" in a loving relationship, or by "affection" and "love" for children, then we will be, based on this superficial concept, creating countless emotional problems for ourselves and setting very strict limits for the quality of our relationships.

Love as an emotion is really fleeting, for all emotions have an activation period and a relatively rapid deactivation compared to thoughts, and even less in relation to values. These (and their motivations associated) thoughts (which researchers designate as cognition) and awareness are more complex and profound human concepts compared to emotions, which are also shared by animals. We can study the origins of emotions, altruism and cooperation in animals, but in human beings, these concepts are more complex because we are a more evolved species. Perhaps it is not so much a question of greater complexity but at least one of greater differentiation toward animals in this respect.

In human beings, there is a much greater evolutionary potential than love only as a short-term emotion, conveyed in culture so frequently, trivialized and superficial. If we overcome the barrier of superficiality, we begin to become aware that love is a concept that needs to relate to deeper and longer-lasting cognition and values in order to promote human beings in a balanced and stable way in relation to sustainable happiness and to substantially improve the quality of relationships.

Eastern doctrines cannot separate the concept of love from the concept of wisdom. If love is considered just an emotion of tenderness, this opens the way to a huge human emotional frailty; therefore, if love is not accompanied by loyalty, respect, intelligence, responsibility, a genuine concern for the development of the other and the care of seeing a relationship as an instrument of progress if it does not base its foundations on intrinsic values, then all this superficial "feeling" is subject to being annulled by other human imperfections, and in relationships, it will always cause great disappointment and suffering.

How can you talk about love when you betray someone else the next day? How one can speak of love if, at the first sign of contradiction, we direct attacks of anger to the person we claim to love? How can we talk about love if, often, what we feel is a set of selfish illusions that, by not being satisfied, we love one person one day and we start loving another the next day? Do we really love a child when we praise the child too much, creating difficulties for the child to realize their own moral deficiencies, instigating in them an exacerbated vanity to have the illusion of "being the smartest in school or the neighborhood," without being focused on the real process of development and evolution of the self, based on the evolution of the child's real self?

Does loving a child mean to just share "positive emotions" in being "your friend" and never seek to assume the difficult mission to stimulate their values, preparing them for a life without illusions of the ego? Even if it means fighting the child's weaknesses with awareness of the real self? Does loving a friend mean just meeting in the clubs for fun or in the cafés to enjoy "pleasant conversation," or does it mean to develop a deeper relationship of genuine concern for the other, to help your friend in their needs and self-evolution, within the limits of one's own emotional, moral, cognitive and material resources?

These honest questions, which we must ask ourselves about love, automatically increase our emotional self-awareness about how societies tend to characterize love superficially in literature, communication, movies, theater and daily life. Clearly, that love became something trivialized and superficial. It is just a matter of level of analysis. If we reflect a little more deeply on the issue of love within relationships, and we ask only some of the questions cited, we find that to truly love someone, we need to improve ourselves in the first place. And a deeper relationship between a couple, between father and son, between friends, between student and teacher, in short, all relationships, constitutes stimuli for our personal progress, but it is necessary motivation for this, in addition to cultivating this value in our lives. This will certainly bring great benefits to our intrapersonal and emotional life and will substantially improve the quality of our own interpersonal relationships.

Everyone we relate to will benefit from someone looking for daily improvement of their real self at the moral, emotional and mental (cognitive) levels. So, before we get into a relationship, we need to develop our real self first. And we always have a lot to learn in our relationships with others, if we are predisposed to it. In the same way that a student can learn a lot from a teacher, a teacher, if they want to, can learn from their students, as long as they have an intrinsic motivation to do so.

Researchers call this a "growth mindset" (Yeager et al., 2019; Dweck, 2008). Humans have an extraordinary potential. If you are willing to learn with intrinsic motivation, you will always be learning.

If, throughout their life, a person is not encouraged to have as motivation and value the intrinsic moral dimension of life, these values will be considered secondary. But if you understand and internalize the intrapersonal and interpersonal benefits of intrinsic morality, assimilate in your real self the essence of morality, certainly you will be much happier emotionally and in relationships. If the growth mindset becomes an intrinsic value throughout a lifetime, it will come from this psychological maturation for the individual at the intrapersonal level, and the quality and selectivity of relationships will be much higher, establishing what researchers call a high-quality connection. On the other hand, if the values of the individual are superficial, full of fantasies and illusions of the ego, the same experiences will not provide psychological maturation, and the individual will continue to suffer from the same mistakes as their personal choices and maintain relationships that are often unhealthy from a psychological point of view.

The connection of the emotion of love to an intrinsic value of growth mindset and personal development, and not of satisfaction of fantasies or caprices of the ego, allows love to be transformed into something deep and lead to mature relationships with a higher level of intimacy and quality. Automatically, this causes humans to be more selective in the choices of their relationships, looking for those who bring evolution and a more genuine and profound love. Certainly, this posture in the face of oneself, relationships and life will result in the presence of more frequent positive emotions and greater emotional maturity and stability for dealing with negative emotions and the difficulties of relationships.

Researchers scientifically define the concept of value as a motivational belief that guides and influences the decisions of human beings in some direction, closely related to the content of the belief in question (Schwartz, 1992).

For example, if the personal development value is very important for the self, then the choices will be influenced by this value. The more internalized this value is in the self, the most influence it will have on intrapersonal life as well as on the choices of people for relationships and the quality of love. Love will be accompanied by the value of personal progress and the other's progress. So, love changes from a short-term emotion to a belief related to the content of the value that cultivates the self and the other. According to some researchers, the connection between an emotion and a belief is designated as sentiment (Frijda et al., 1991; Oatley, 2000).

The Difference Between the Concepts of Emotion and Sentiment in the Context of Relationships

In this context, sentiment is a predisposition resulting from the combination of short-term emotions (basic, moral and positive emotions, which are, by nature, a short-term phenomenon) with values and motivational beliefs (Frijda et al., 1991; Oatley, 2000).

As a connection between basic and moral emotions with higher-order cognitions – motivational values and beliefs – the concept of sentiment has some advantages compared to the concept of emotions: It is uniquely human, not shared by other animals; it is a longer-lasting concept for humans compared to emotions; it allows greater stability in time when compared to emotions because it relates to higher-order cognitions (values, beliefs and motivations); it is less likely to be diluted due to daily emotional conflicts (less "fragile" and unstable than emotions because sentiments are much less context dependent and require less self-control efforts); and it makes no sense to think about "sentimental self-control" but of connecting the emotions generated in contexts specific to the intrinsic values applied to each situation. (We call this the situational reappraisal process of intrinsic values.)

This will make human beings capable of applying intrinsic values to each situation and, consequently, assign these values priority over the basic and moral emotions, which are short term and conducive to impulsivity. In this way, the values and sentiments resulting will guide the cognitive reappraisal of the situation and, consequently, will have a greater chance of success in emotional regulation compared to merely trying to inhibit negative emotions.

This concept of sentiment is quite different from the concept proposed by Professor Antonio Damasio. For Damasio, feeling originally emerges as a subjective reflection of the state of the body and is not related to any motivational belief, only with sensations of the body. The purpose of Damasio's work was to identify the neural origins of emotion and its connection to the brain (Damasio, 2005).

The concept of sentiment proposed in this book is very different. What was proposed here was to identify and understand long-term emotions, connected with values and beliefs, also called, by some researchers, sentiments (Frijda et al., 1991; Oatley, 2000).

There is virtually no line of scientific research that systematically studies the concept of sentiment with its neural bases, what we can call the neural bases of sentiments and their connection to values (Nussbaum, 2003).

This is of crucial importance to the study of human relationships, particularly for such a long-term and stable concept as sentiment. Much of our happiness or unhappiness, of our emotional life, our progress or our personal ruin is highly influenced by the quality of our long-term relationships and our sentiments. The in-depth scientific inquiry into the characteristics of these sentiments through the interplay between emotions and long-term values and beliefs and its consequences to relationships and forms of personal adjustment to each of them, since each one constitutes a single system, is an important exercise of emotional intelligence applied to relationships.

References

Arueti, M., Perach-Barzilay, N., Tsoory, M. M., Berger, B., Getter, N., & Shamay-Tsoory, S. G. (2013). When two become one: The role of oxytocin in interpersonal coordination and cooperation. *Journal of Cognitive Neuroscience, 25*(9), 1418–1427. https://doi.org/10.1162/jocn_a_00400

Bartz, J. A., Zaki, J., Bolger, N., & Ochsner, K. N. (2011). Social effects of oxytocin in humans: Context and person matter. *Trends in Cognitive Sciences, 15*(7), 301–309. https://doi.org/10.1016/j.tics.2011.05.002

Bauman, Z. (2001): *The individualized society*. Polity.

Bauman, Z. (2003). *Liquid love: On the frailty of human bonds*. Polity.

Blais, M. R., Sabourin S., Boucher, C., &Vallerand, R. J. (1990). Toward a motivational model of couple happiness. *Journal of Personality and Social Psychology*, *59*, 1021–1031. https://doi.org/10.1037/0022-3514.59.5.1021

Cacioppo, J. T., & Patrick, W. (2008). *Loneliness: Human nature and the need for social connection*. WW Norton & Company.

Damasio, A. R. (2005). *Descartes' error: Emotion, reason, and the human brain*. Penguin.

De Dreu, C. K., Greer, L. L., Handgraaf, M. J., Shalvi, S., Van Kleef, G. A., Baas, M., Velden, F. S. T., Dijk, E. V., & Feith, S. W. W. (2010). The neuropeptide oxytocin regulates parochial altruism in intergroup conflict among humans. *Science*, *328*(5984), 1408–1411. https://doi.org/10.1126/science.1189047

De Dreu, C. K., Greer, L. L., Van Kleef, G. A., Shalvi, S., & Handgraaf, M. J. (2011). Oxytocin promotes human ethnocentrism. *Proceedings of the National Academy of Sciences*, *108*(4), 1262–1266. https://doi.org/10.1073/pnas.1015316108

de Waal, F. (2016). *Are we smart enough to know how smart animals are?* WW Norton & Company.

Domes, G., Heinrichs, M., Michel, A., Berger, C., & Herpertz, S. C. (2007). Oxytocin improves "mind-reading" in humans. *Biological Psychiatry*, *61*(6), 731–733. https://doi.org/10.1016/j.biopsych.2006.07.015

Duque-Wilckens, N., Steinman, M. Q., Busnelli, M., Chini, B., Yokoyama, S., Pham, M., Laredo, S. A., Hao, R., Perkeybile, A. M., Minie, V. A., Tan, P. B., Bales, K. L., & Trainor, B. C. (2018). Oxytocin receptors in the anteromedial bed nucleus of the stria terminalis promote stress-induced social avoidance in female California mice. *Biological Psychiatry*, *83*(3), 203–213. https://doi.org/10.1016/j.biopsych.2017.08.024

Dweck, C. S. (2008). *Mindset: The new psychology of success*. Random House Digital, Inc.

Eckstein, M., Scheele, D., Weber, K., Stoffel-Wagner, B., Maier, W., & Hurlemann, R. (2014). Oxytocin facilitates the sensation of social stress. *Human Brain Mapping*, *35*(9), 4741–4750. https://doi.org/10.1002/hbm.22508

Frijda, N. H., Mesquita, B., Sonnemans, J., & Van Goozen, S. (1991). The duration of affective phenomena or emotions, sentiments, and passions. In K. T. Strongman (Ed.), *International review of studies on emotion, volume 1* (pp. 187–225). John Wiley & Sons.

Fromm, E. (1956). *The art of loving*. Harper & Row.

Guastella, A. J., Mitchell, P. B., & Dadds, M. R. (2008). Oxytocin increases gaze to the eye region of human faces. *Biological Psychiatry*, *63*(1), 3–5. https://doi.org/10.1016/j.biopsych.2007.06.026

Heinrichs, M., Baumgartner, T., Kirschbaum, C., & Ehlert, U. (2003). Social support and oxytocin interact to suppress cortisol and subjective responses to psychosocial stress. *Biological Psychiatry*, *54*(12), 1389–1398. https://doi.org/10.1016/S0006-3223(03)00465-7

Hurlemann, R., Patin, A., Onur, O. A., Cohen, M. X., Baumgartner, T., Metzler, S., Dziobek, I., Gallinat, J., Wagner, M., Maier, W., & Kendrick, K. M. (2010). Oxytocin enhances amygdala-dependent, socially reinforced learning and emotional empathy in humans. *Journal of Neuroscience*, *30*(14), 4999–5007. https://doi.org/10.1523/JNEUROSCI.5538-09.2010

Kasser, T. (2016). Materialistic values and goals. *Annual Review of Psychology*, *67*, 489–514. https://doi.org/10.1146/annurev-psych-122414-033344

Kosfeld, M., Heinrichs, M., Zak, P. J., Fischbacher, U., & Fehr, E. (2005). Oxytocin increases trust in humans. *Nature*, *435*(7042), 673–676. https://doi.org/10.1038/nature03701

Lee, H. J., Macbeth, A. H., Pagani, J. H., & Young 3rd, W. S. (2009). Oxytocin: The great facilitator of life. *Progress in Neurobiology, 88*(2), 127–151. https://doi.org/10.1016/j.pneurobio.2009.04.001

Liu, Y., Li, S., Lin, W., Li, W., Yan, X., Wang, X., Pan, X., Rutledge, R. B., & Ma, Y. (2019). Oxytocin modulates social value representations in the amygdala. *Nature Neuroscience, 22*(4), 633–641. https://doi.org/10.1038/s41593-019-03511

McPherson, M., Smith-Lovin, L., & Brashears, M. E. (2006). Social isolation in America: Changes in core discussion networks over two decades. *American Sociological Review, 71*(3), 353–375. https://doi.org/10.1177/000312240607100301

Myers, D. G. (2001). *The American paradox: Spiritual hunger in an age of plenty*. Yale University Press.

Nussbaum, M. C. (2003). *Upheavals of thought: The intelligence of emotions*. Cambridge University Press.

Oatley, K. (2000). The sentiments and beliefs of distributed cognition. In N. H. Frijda, A. S. R. Manstead, & S. Bem (Eds.), *Studies in emotion and social interaction. Emotions and belief: How feelings influence thoughts* (pp. 78–107). Cambridge University Press.

Ryan, R. M., Stiller, J., & Lynch, J. H. (1994). Representations of relationships to teachers, parents, and friends as predictors of academic motivation and self-esteem. *Journal of Early Adolescence, 14*, 226–249. https://doi.org/10.1177/027243169401400207

Scheele, D., Striepens, N., Güntürkün, O., Deutschländer, S., Maier, W., Kendrick, K. M., & Hurlemann, R. (2012). Oxytocin modulates social distance between males and females. *Journal of Neuroscience, 32*(46), 16074–16079. https://doi.org/10.1523/JNEUROSCI.2755-12.2012

Schwartz, S. H. (1992). Universals in the content and structure of values: Theoretical advances and empirical tests in 20 countries. *Advances in Experimental Social Psychology, 25*(1), 1–65. https://doi.org/10.1016/S0065-2601(08)60281-6

Shamay-Tsoory, S. G., & Abu-Akel, A. (2016). The social salience hypothesis of oxytocin. *Biological Psychiatry, 79*(3), 194–202. https://doi.org/10.1016/j.biopsych.2015.07.020

Spengler, F. B., Scheele, D., Marsh, N., Kofferath, C., Flach, A., Schwarz, S., Stoffel-Wagner, B., Maier, W., & Hurlemann, R. (2017). Oxytocin facilitates reciprocity in social communication. *Social Cognitive and Affective Neuroscience, 12*(8), 1325–1333. https://doi.org/10.1093/scan/nsx061

Stevenson, B., & Wolfers, J. (2007). Marriage and divorce: Changes and their driving forces. *Journal of Economic Perspectives, 21*(2), 27–52. https://doi.org/10.1257/jep.21.2.27

Van Anders, S. M., Goodson, J. L., & Kingsbury, M. A. (2013). Beyond "oxytocin= good": Neural complexities and the flipside of social bonds. *Archives of Sexual Behavior, 42*(7), 1115–1118. https://doi.org/10.1007/s10508-013-0134-9

Vandewater, E. A., Bickham, D. S., & Lee, J. H. (2006). Time well spent? Relating television use to children's free-time activities. *Pediatrics, 117*(2), e181–e191. https://doi.org/10.1542/peds.2005-0812

Yeager, D. S., Hanselman, P., Walton, G. M., Murray, J. S., Crosnoe, R., Muller, C., Tipton, E., Schneider, B., Hulleman, C. S., Hinojosa, C. P., Paunesku, D., Romero, C., Flint, K., Roberts, A., Trott, J., Iachan, R., Buontempo, J., Yang, S. M., Carvalho, C. M., . . ., Dweck, C. S. (2019). A national experiment reveals where a growth mindset improves achievement. *Nature, 573*(7774), 364–369. https://doi.org/10.1038/s41586-019-1466-y

Zahn, R., de Oliveira-Souza, R., & Moll, J. (2020). Moral motivation and the basal forebrain. *Neuroscience & Biobehavioral Reviews, 108*, 207–217. https://doi.org/10.1016/j.neubiorev.2019.10.022

Part II

Emotional Intelligence Applied to Education

Chapter 7

Educational Programs in Emotional Intelligence

Some scientific studies have shown that the education of emotional intelligence in schools has had several positive and promising effects on the lives of children and adolescents, including a higher rate of integration into higher education, better jobs, better mental and emotional health and a reduction in crime and drug use (Jones et al., 2015; Taylor et al., 2017).

The first major study on the effects of emotional intelligence education on children and adolescents involved the impressive number of 270,000 participants, evidencing a significant increase in academic success (an increase in 11 percentage points in the North American standardized academic tests compared to students who did not have an education in emotional intelligence) as well as a greater capacity of children and adolescents to manage stress and depression (Durlak et al., 2011).

The CASEL Education Program of Social and Emotional Learning

CASEL (Collaborative for Academic, Social, and Emotional Learning), from the University of Illinois, led by investigator Ross Weissberg, was one of the first initiatives in the United States dedicated to children's and adolescents' social and emotional learning (SEL). The latest meta-analysis, a statistical analysis of the results of several scientific studies of the impact of emotional intelligence education on children and adolescents demonstrates very positive results (Taylor et al., 2017).

They evaluated more than 97,000 children and adolescents, from kindergarten to secondary school. Interventions in SEL, in which emotional intelligence skills were taught, have shown a clear improvement in the attitudes of children and adolescents toward self, others and school, as well as the production of positive behaviors in the school environment, better school performance, fewer behavioral problems, fewer negative and emotional problems and lower drug abuse among children and adolescents (Taylor et al., 2017).

According to CASEL, SEL encompasses five groups of emotional intelligence competencies: self-awareness, self-management, social awareness, relationship management and responsible decision-making (CASEL, 2020).

Self-Awareness

Self-awareness, according to the CASEL model, is a group of competencies whose objective is to recognize in a precise way emotions, thoughts and values, and how they influence behavior. It also encompasses accurately identifying and assessing strengths and limitations, in a sense of confidence and optimism, and from a perspective of growth mindset and personal growth. Self-awareness includes the following competencies:

Emotion identification – Recognize what emotion you are feeling, whether basic or moral emotions. Being able to identify the emotion accurately requires a detailed knowledge of each emotion that is described in the beginning of this book, allowing an increase in self-awareness of the patterns of emotions generated in the most varied situations.

Accurate emotional self-awareness – Discern accurately which emotion the person is feeling during a specific situation. Emotional identification is the first step for accurate self-awareness, but self-awareness requires attention to the emotions the self is feeling during the situation itself. It requires a lot of mindful attention inward during the situation. The more the child or adolescent is aware of an emotion within a situation, it increases the probability of control of this emotion during the situation.

Recognize one's own strengths and limitations – realistically reflect on talents, virtues and skills, as well as on defects, difficulties and weaknesses in various situations.

Self-confidence – Trust oneself; be centered on virtues and abilities to allow personal progress centered on the self. It is important to distinguish between trusting the self and trusting the "ego." Trust in the self comes from an intrinsic motivation, centered on personal progress, and promotes persistence and resilience, while trust in the ego only generates a narcissistic attitude.

Self-efficacy – This concept, created by Bandura (1982), is composed of two elements: the belief that behavior can lead to a certain result and the individual's belief that they can produce this behavior. Several studies have shown that the greater the self-efficacy, the greater is the individual's achievement capacity.

Self-Management

Self-management is the ability to manage one's own emotions, thoughts and behaviors adequately in different situations, and it comprises the following competencies:

Emotional self-control – Regulate an emotion in the face of a situation. This is considered the most important of emotional intelligence skills. Increased emotional self-awareness during a situation increases the probability of emotional self-control. It is the time between the emotional impulse and the behavior during the situation, influenced by this emotion. Thus, increasing self-control is the ability to expand the time of cognitive control of emotion before moving to action. Greater emotional self-control in children predicts academic success and productivity in the classroom and on standardized tests, even controlling the level of quantitative intelligence quotient or IQ (Graziano et al., 2007).

Stress management – Manage emotional pressures intelligently in various situations. For good stress management, it is important to develop emotional self-awareness and self-control during a stressful situation.

Self-discipline – This involves organizing one's own mind, emotions and behaviors to create productive patterns of behavior and mental and emotional well-being.

Intrinsic motivation – This is the motivation derived from the self. Hundreds of studies have shown that over time, intrinsic motivation is the best form of individual sustainable motivation, and it has benefits in academic success, psychological well-being and happiness, relationships, goal setting, sport and exercise, education, self-esteem and work, among many other areas (Ryan & Deci, 2000).

Setting goals – This involves setting objectives in a SMART way, i.e., in a specific, measurable, achievable, time-defined way (Doran, 1981). This covers the development of an action plan with the aim of guiding and motivating the adolescent to achieve a goal (Locke et al., 2002). Several studies demonstrated that setting goals leads to an increase in performance and satisfaction (Locke et al., 2002). Another study group revealed that the more goals that are set related to intrinsic aspirations (i.e., personal development, relationships with deep meaning and/or that contribute to a better world), the higher will be individual levels of well-being. On the other hand, the more the goals are related to extrinsic aspirations (fame, wealth and physical beauty), the higher are the levels of dissatisfaction, depression and anxiety (Sheldon & Kasser, 1998; Kasser & Ryan, 1993, 1996).

Organizational skills – These are competencies that children and adolescents should develop to achieve success within an organization, such as primary, secondary and higher school as well as companies when adults. Two examples of organizational skills are organizational citizenship behaviors and organizational awareness.

The former is a set of positive behaviors freely chosen by the child and adolescent with the aim of contributing to the organization to which they are affiliated. These behaviors are related to the sense of community, of belonging of the individual to an organization, and the individual's contribution to the organization's mission (Graham, 1991; van Dyne et al., 1994).

Organizational awareness – The goal is to become aware of the mission of the organization and its role in society and to understand what the individual's contribution may be for the mission of the organization. In addition, organizational awareness involves understanding how the organization works formally and informally to help the individual position themself in it effectively and positively (Spitzer et al., 2015).

Social Awareness

Social awareness is the capacity to take the perspective of others, including people from diverse cultures, academic backgrounds and social classes. It means being able to understand and internalize the social and ethical norms of social life and family, community and school as sources of social support. It is understanding the following competencies:

Perspective taking – This is cognitively assuming the perspective of other people, to put themselves mentally in their place and in their situation, and to understand what they feel and think about in a particular situation. This competence of emotional intelligence is also called cognitive empathy and is connected to what researchers call theory of mind (Shamay-Tsoory et al., 2009).

Empathy – This is the capacity to feel what others are feeling. This competence is called emotional empathy (Shamay-Tsoory et al., 2009). Several scientific studies have demonstrated the relationship between empathy and altruism (Batson et al., 1991). Thus, the more the children develop the competence of emotional empathy, the more they will tend to be benevolent, generous and altruistic.

Appreciation of diversity – This is shown when identifying and appreciating differences in beliefs and social and economic classes of the various individuals and social groups. To appreciate means to recognize positive points in all these differences and verbalize them in interpersonal relationships, to stimulate interpersonal constructive and positive relationships.

Respect for others – Respect the other as they are, without judgment of behaviors or interpersonal or social rejection.

Relationship Management

Relationship management is the set of competencies related to the creation and maintenance of constructive and healthy relationships with people and groups from different cultures and social classes. It is to be able to communicate clearly, to listen actively and generatively, to cooperate with others, to resist inappropriate social pressures, to negotiate and manage conflicts constructively and to try to help those in need. It encompasses the following competencies:

Communication – Communicate effectively and constructively with other people to produce meaningful relationships.

Active social involvement – Proactively take the initiative in each social goal or mission.

Building meaningful relationships – Manage interpersonal relationships with deep meaning based on intrinsic values of development and mutual assistance and of affection and sharing of financial, emotional, social and knowledge resources.

Teamwork – Establish positive relationships with other team members as well as contribute significantly to achieving their goals.

Responsible Decision-Making

Responsible decision-making is the capacity to make choices about personal behaviors and social interactions based on ethics and morality, with concerns about the safety of others and according to appropriate social norms. It involves a realistic assessment of the consequences of the various actions and consideration for the well-being of oneself and others.

It comprises the following competencies:

Identification of problems – Recognize the problems inherent in a situation in a precise and objective way.
Analysis of situations – Analyze situations broadly and objectively, above one's own personal perspectives, and take on the various possible perspectives existing in each situation.
Problem-solving – Propose effective solutions for the most varied types of problems.
Evaluation – Assess the impact of the consequences of a given decision in various contexts.
Reflection – Reflect in a timely manner on the causes and consequences of decisions to allow responsible decision-making. This competence enables a reduction of impulsiveness in making decisions that would generate subsequent repentance of the child and the adolescent, and adults.
Ethical and moral responsibility – There is moral engagement inherent in a particular decision. Most of the problems worldwide are a consequence of the absence of ethical and moral responsibility and moral disengagement processes prior to decision-making (Bandura, 2002, 2016; Detert et al., 2008). A scientific study involving about 17,000 children and adolescents demonstrated a relationship between moral disengagement and aggressive behavior (Gini et al., 2014). Therefore, it is important to teach children and adolescents to engage morally in their decisions.

The return on investment of an emotional intelligence education program is 11 to 1 – that is, for every dollar invested, the return is $11!

This is very important for governments that finance projects of this nature in public schools as well as for private schools that also wish to follow this path. In addition to the positive psychological and emotional effects and the financial feasibility, there is the benefit of reducing poverty and increasing the social and economic mobility of children and adolescents when these projects are intended for less-favored socio-economic classes (AEI/Brookings Working Institute, Working Group on Poverty and Opportunity, 2015). For example, the American Enterprise Institute and the Brookings Institution strongly recommend the learning of emotional intelligence to help reduce poverty and increase the economic and social mobility of children and adolescents (AEI/Brookings Working Institute, Working Group on Poverty and Opportunity, 2015).

CASEL's intervention model not only covers the education of children and adolescents in emotional intelligence in the classroom context, it also emphasizes the importance of involving all school staff and teacher training as key pieces for the success of emotional intelligence education in school practices and policies at the organizational level or larger systems level as well as involving families in the process of learning emotional intelligence as an instrument for strengthening these skills in the day-to-day lives of children and adolescents. In addition, the intervention needs to be able to create a community-like school environment, to strengthen and enhance the emotional and social skills at school and in the family. The success of a SEL program depends on how it is implemented in the school, i.e., the degree of involvement of its employees, the motivation and dedication of teachers and parents and the collaboration and active involvement of children and adolescents throughout this process.

CASEL is not the only emotional intelligence education program – additional programs are discussed next.

The RULER Program of Emotional Intelligence Education (Yale University)

Another program for teaching emotional intelligence to children and adolescents is the RULER (R, recognize; U, understand; L, labeling, naming; E, expressing; and R, regulating, regular) method.

Thus, according to this approach, the learning of emotional intelligence involves the ability to recognize emotion at the level of perception, to understand emotion and its effect on thought, to specifically name emotions, to express emotions appropriately and to adjust this emotion in social situations.

The RULER approach is a practical version of application of Mayer and Salovey's emotional intelligence model to the education of children and adolescents (Mayer & Salovey, 1997).

The Mayer and Salovey model (1997) on which the RULER method is based is composed of four parts, each of which represents a set of capacities, ordered by complexity. Thus, the acquisition of an aptitude depends on the acquisition of the previous suitability provided for in the model. Next, you analyze that model.

The first stage of the Mayer and Salovey (1997) model for learning emotional intelligence is perception, evaluation and expression of emotion – become aware of and understand what emotion you are feeling, assess the emotion functionality or dysfunction and express the emotion appropriately.

The second stage of the model is the ability to understand that emotions are resources that can facilitate or hinder thought processes. For example, joy can facilitate a creative mental process, while fear can block it. The main capabilities associated with this stage are the ability to generate or trigger emotions to facilitate memories of events or judgment about something and the ability to identify which specific emotions facilitate and which ones make it difficult to think.

The third stage of learning emotional intelligence is the ability to understand complex questions about different emotions; understand the causes and consequences of emotions; understand complex emotions, such as ambivalence; and transition from dysfunctional negative emotions to positive and constructive emotions for individuals and relationships.

The fourth and final stage of learning emotional intelligence constitutes the most difficult – the ability to regulate emotions; be open to emotions, both pleasure as well as pain; monitor and reflect on emotional impulses; generate prolonged emotions or disconnect from an emotional state; and generate functional and positive emotions on oneself and others.

In this last stage, individuals can place emotional information in action in an appropriate and intelligent manner; be open to emotional experience and practice to engage in behavior generators of positive and intrinsically rewarding emotions; regulate mood to maintain positive emotions they are feeling and to modify the negative emotional state they may have been feeling, using several strategies, including avoiding non-pleasurable activities or engaging in activities they know to be intrinsically rewarding; and acquire knowledge of the causes and consequences of their emotional experiences and apply them correctly in several situations, whether negative or positive emotions (Izard et al., 2008).

An important line of scientific research is the understanding of how parents and educators help children and adolescents regulate their emotions. There is a clear distinction among educators who assume a mentoring stance of children in the regulation of their emotions and educators who do not consider the emotions of children (Gottman et al., 1997). In the case of the former, they treat children's emotions as a useful source of information and send the message to children that their emotions are accepted.

Thus, these educators become aware of children's emotions, identify their negative emotions, and help them become aware of these emotions and name them, in a process called emotional literacy.

This last part is an important step that precedes emotional self-control. In turn, children helped by educators in a systematic way learn to have more emotional self-awareness of their emotions, as they feel accepted by parents and educators, and learn to face their patterns of emotions autonomously and thus tend to feel a higher level of self-acceptance. In addition, they learn, with the help of educators (who assume the role of mentors), to reflect on these emotions and the negative thoughts that generate them, and to cognitively reappraise in each situation, taking into account a system of values and motivations important to the self and surrounded by psychological safety to put in practice the intrinsic values important for the self, and the values of the family and the community in which they are inserted.

The strategy in which educators assume the role of children's mentors in regulating their emotions has very positive results in the emotional regulation of children, elevating their capacity for emotional intelligence, as well as in various other areas, such as improving academic performance, controlling impulsivity, and increasing happiness and emotional well-being (Dunsmore et al., 2013; Gottman et al., 1997).

On the other hand, educators who do not consider children's emotions assume opposite posture, i.e., communicate to children or students that certain emotions are unacceptable.

Specifically, this type of educator is not aware of children's emotions, uses an impoverished emotional vocabulary, and tries, forcefully, to impose a change, rather than helping the child to take responsibility in the process of emotional regulation. Consequently, the child feels that their negative emotions are not accepted and therefore may develop a lower self-acceptance, which causes the child to suppress negative emotions, which entails emotional problems or is linked to excessive emotional rumination or negative consequences of unresolved negative emotions in themselves.

For educators who do not take into account the regulation of emotions in children, the consequences are worse academic performance, greater difficulty in controlling one's own negative emotions and a lower level of emotional well-being (Dunsmore et al., 2013; Gottman et al., 1997).

Another important aspect of the ability to regulate emotions is the individual can then help to improve the emotions of others or generate positive emotions within the context of relationships. In addition, individuals develop a greater sense of self-efficacy and social value because they can help others feel happier.

The fact is that in addition to applications in the context of an interpersonal relationship, children and adolescents with the ability to regulation emotions are more successful in the sphere of relationship management.

The RULER approach of Yale University's Center for Emotional Intelligence is similar to that carried out by CASEL because in addition to promoting training for teachers and pupils, it also develops training for school staff and children's families, helping the education system create a sense of community learning.

References

AEI/Brookings Institute. Working Group on Poverty and Opportunity (2015). *Opportunity, responsibility, and security: A consensus plan for reducing poverty and restoring the American dream*. AEI.
Bandura, A. (1982). Self-efficacy mechanism in human agency. *American Psychologist*, 37(2), 122–147. https://doi.org/10.1037/0003-066X.37.2.122
Bandura, A. (2002). Selective moral disengagement in the exercise of moral agency. *Journal of Moral Education*, 31(2), 101–119. https://doi.org/10.1080/0305724022014322
Bandura, A. (2016). *Moral disengagement: How people do harm and live with themselves*. Worth Publishers.

Batson, C. D., Batson, J. G., Slingsby, J. K., Harrell, K. L., Peekna, H. M., & Todd, R. M. (1991). Empathic joy and the empathy-altruism hypothesis. *Journal of Personality and Social Psychology, 61*(3), 413–426. https://doi.org/10.1037/0022-3514.61.3.413
CASEL. (2020). Core SEL Competencies [Web page]. https://casel.org/core-competencies/
Detert, J. R., Treviño, L. K., & Sweitzer, V. L. (2008). Moral disengagement in ethical decision making: A study of antecedents and outcomes. *Journal of Applied Psychology, 93*(2), 374. https://doi.org/10.1037/0021-9010.93.2.374
Doran, G. T. (1981). There's a SMART way to write management's goals and objectives. *Management Review, 70*(11), 35–36.
Dunsmore, J. C., Booker, J. A., & Ollendick, T. H. (2013). Parental emotion coaching and child emotion regulation as protective factors for children with oppositional defiant disorder. *Social Development, 22,* 444–466. https://doi.org/10.1111/j.1467-9507.2011.00652.x
Durlak, J. A., Weissberg, R. P., Dymnicki, A. B., Taylor, R. D., & Schellinger, K. B. (2011). The impact of enhancing students' social and emotional learning: A meta-analysis of school-based universal interventions. *Child Development, 82*(1), 405–432. https://doi.org/10.1111/j.1467-8624.2010.01564.x
Gini, G., Pozzoli, T., & Hymel, S. (2014). Moral disengagement among children and youth: A meta-analytic review of links to aggressive behavior. *Aggressive Behavior, 40*(1), 56–68. https://doi.org/10.1002/ab.21502
Gottman, J. M., Katz, L. F., & Hooven, C. (1997). *Meta-emotion: How families communicate emotionally*. Lawrence Erlbaum Associates.
Graham, J. W. (1991). An essay on organizational citizenship behavior. *Employee Responsibilities and Rights Journal, 4*(4), 249–270. https://doi.org/10.1007/BF01385031
Graziano, P. A., Reavis, R. D., Keane, S. P., & Calkins, S. D. (2007). The role of emotion regulation and children's early academic success. *Journal of School Psychology, 45*(1), 3–19. https://doi.org/10.1016/j.jsp.2006.09.002
Izard, C., Stark, K., Trentacosta, C., & Schultz, D. (2008). Beyond emotion regulation: Emotion utilization and adaptive functioning. *Child Development Perspectives, 2*(3), 156–163. http://dx.doi.org/10.1111/j.1750-8606.2008.00058.x
Jones, D. E., Greenberg, M., & Crowley, M. (2015). Early social-emotional functioning and public health: The relationship between kindergarten social competence and future wellness. *American Journal of Public Health, 105*(11), 2283–2290. https://doi.org/10.2105/AJPH.2015.302630
Kasser, T., & Ryan, R. M. (1993). A dark side of the American dream: Correlates of financial success as a central life aspiration. *Journal of Personality and Social Psychology, 65,* 410–422. https://doi.org/10.1037/0022-3514.65.2.410
Kasser, T., & Ryan, R. M. (1996). Further examining the American dream: Differential correlates of intrinsic and extrinsic goals. *Personality and Social Psychology Bulletin, 22,* 280–287. https://doi.org/10.1177/0146167296223006
Locke, Edwin A., & Latham, Gary P. (2002). Building a practically useful theory of goal setting and task motivation: A 35-year odyssey. *American Psychologist, 57*(9),705–717. https://doi.org/10.1037/0003-066x.57.9.705
Mayer, J. D., & Salovey, P. (1997). What is emotional intelligence? In P. Salovey & D. Sluyter (Eds.), *Emotional development and emotional intelligence: Educational implications* (pp. 3–31). Basic Books.
Ryan, R. M., & Deci, E. L. (2000). Self-determination theory and the facilitation of intrinsic motivation, social development, and well-being. *American Psychologist, 55*(1), 68–78. https://doi.org/10.1037/0003-066x.55.1.68
Shamay-Tsoory, S. G., Aharon-Peretz, J., & Perry, D. (2009). Two systems for empathy: A double dissociation between emotional and cognitive empathy in inferior frontal

gyrus versus ventromedial prefrontal lesions. *Brain, 132*(3), 617–627. https://doi.org/10.1093/brain/awn279

Sheldon, K. M., & Kasser, T. (1998). Pursuing personal goals: Skills enable progress but not all progress is beneficial. *Personality and Social Psychology Bulletin, 24*, 1319–1331. https://doi.org/10.1177/01461672982412006

Spitzer, W., Silverman, E., & Allen, K. (2015). From organizational awareness to organizational competency in health care social work: The importance of formulating a "profession-in-environment" fit. *Social Work in Health Care, 54*(3), 193–211. https://doi.org/10.1080/00981389.2014.990131

Taylor, R. D., Oberle, E., Durlak, J. A., & Weissberg, R. P. (2017). Promoting positive youth development through school-based social and emotional learning interventions: A meta-analysis of follow-up effects. *Child Development, 88*(4), 1156–1171. https://doi.org/10.1111/cdev.12864

van Dyne, L., Graham, J. W., & Dienesch, R. M. (1994). Organizational citizenship behavior: Construct redefinition, measurement, and validation. *Academy of Management Journal, 37*(4), 765–802. https://doi.org/10.5465/256600

Chapter 8

Dimensions of Focus of Emotional Intelligence

Self and Others

Daniel Goleman, the world's largest promoter of the concept of emotional intelligence, wrote, together with Peter Senge, a senior lecturer at the Massachusetts Institute of Technology (MIT), a book titled *The Triple Focus*, in which they advocate a new approach to the education of emotional intelligence in schools (Goleman & Senge, 2014). This book points to three major focuses: (1) self, (2) others and (3) large systems (Goleman & Senge, 2014).

Due to space limitations and because this is not the focus of this book, we do not approach the third focus on large systems, because it would require a whole framework on the science of systems thinking developed at MIT that gave rise to the concept of sustainability and systems thinking applied to management, climate change, economics, business and education. For the reader interested in deepening their knowledge in the third focus, see Goleman & Senge (2014).

The focus in this book is on the self dimension of emotional intelligence as well as the application of emotional intelligence in relationships with others.

Self

Emotional intelligence education teaches children and adolescents to deal with their selves, their own emotions, their values, their motivations and their inner thoughts and strengths, helping them to build their personal identities. It is similar to the concept of intrapersonal intelligence (Gardner, 2011).

One of the first learning stages of emotional intelligence is called emotional literacy, which is the capacity of children and adolescents to name accurately the emotion (specific basic or moral emotion) they are feeling at a certain time (Cohen, 2001).

For example, some interventions demonstrated that many children tended to eat in excess when they were sad, and they could not identify the relationship between the fact that they felt sad and the urge to eat too much. The first step of education in emotional intelligence is to help them to be able to identify accurately which emotion they were feeling at a certain moment (emotional self-awareness) and the connection of an emotion (e.g., in this

DOI: 10.4324/9781003503880-10

case, sadness) and an action tendency (e.g., in this case, eating too much) that is related to this specific emotion.

After being able to identify the emotion in themselves at a given time, the child should be encouraged to learn practical strategies to regulate this emotion (emotional self-control) before going to the action tendency to eat too much.

Thus, several exercises are carried out in the classroom with the aim of stimulating children to reflect on the situations and emotions that provoke them, and to develop practical strategies to adequately regulate these emotions. These exercises serve as a stimulus for the practice of emotional self-awareness throughout the day.

Children also learn to know the types of basic emotions and how they manifest themselves in their mind and in their body, particularly in facial expressions and body language, and learn to regulate it, according to their nature. The animated movie *Inside Out*, produced by Pixar animation studios and released by Walt Disney Pictures in 2015, illustrates to children the basic emotions and how they manifest themselves, their behavioral effects and how they influence thoughts.

Through these exercises, children are encouraged to increase their level of self-awareness of their thoughts and how they influence emotions, and how they influence, in turn, emotions and behaviors through the habit of thinking and reflecting on the patterns of emotions and thoughts they tend to have and to identify the internal and external influences that generate them.

Another important aspect of the self is the ability to reflect on what are the most important values of their lives, also called self-central values. The stimulus to systematic reflection about these values that are more central to the sense of self helps them build their identity, motivations and intrinsic values. The more training they receive in this area, the more children can clarify and strengthen their values, motivations and personal identity. They will develop the capacity to differentiate their identities from the identities of others, and this will help, as a consequence, to create a more stable and balanced self, from a cognitive and emotional perspective.

To know oneself, with strengths and limitations, values and preferences, helps you define your identity and become less susceptible to negative external influences that may cause inner confusion in your self and cause emotional, mental or behavioral unbalance.

The self-directed learning model based on emotional intelligence (Boyatzis, 2002) is used as a methodology to help in the construction of the self, from a constant personal development and a growth mindset throughout life (Dweck, 2012). Initially proposed in the context of adults (Boyatzis, 2002), this book suggests adapting this model to children and adolescents.

The self-directed learning model developed by Boyatzis (2002) is composed of six discoveries, each of which represents a discontinuity. This model argues that learning requires several resources that should be used at different stages, what he called "discoveries."

The First Discovery: The Construction of the Ideal Self

The first discovery is the search for answers to the following questions that children and teenagers should be encouraged to make: Who is my ideal self? Who do I wish to be (ideal self)? Ideal self is an image constructed by the child and adolescent of whom the self would like to be. This image emerges from the ideal self and should be generated by the child and by the adolescent. From the point of view of psychology, in this discovery of the ideal self, the self is seen as a dynamic process in constant construction of the identity, and that evolves over time, throughout life (Blasi, 1988; Deci & Ryan, 1991; Loevinger, 1976). The ideal self allows children and adolescents to create an intrinsic frame of reference that relies on their intrinsic values and motivations and to learn to deal appropriately with external influences.

The paradox is that in the society in which we live, despite the excess of stimuli and easy access for children and adolescents to technical, cultural and scientific knowledge through the Internet, there is in modern society a great vacuum of human ideals. The fascination with technology and the multiplicity of information available to human beings, on the one hand, make human beings have high expectations regarding the technology but low expectations for the quality of relationships, for ethics, principles and human values (Turkle, 2017, 2003, 1996).

Ideals are mental models that encourage children, adolescents and even adults to seek and wish to be better as human beings. The clearer and more well defined the ideal self, the better is the stimulus to practical good in the real world. The ideal provides a positive stimulus for human achievement, which constitutes the elevation of the human condition to a better condition. If all members of society genuinely seek to be better as human beings, and seek to acquire, through the educational process, not only intellectual and technical development but also emotional, moral and ethical development, and at the level of virtues, which currently have scientific bases (Peterson & Seligman, 2004), certainly over time there will be improvement of individual human beings and the society as a whole. It is of fundamental importance that education become a facilitator that all individuals seek to improve themselves as human beings, and not just a tool to guarantee the acquisition of technical and scientific knowledge.

Thus, the ideal self is a mental model, where the behaviors of children, adolescents and adults are based, oriented. Regarding learning, all the scientific research that has been developed since the 1990s (Salovey & Mayer, 1990) on emotional intelligence allows for emotional and social learning. At the level of virtues, there are currently scientific bases that allow the construction of programs for the development of the virtues of children, adolescents and adults (Peterson & Seligman, 2004; Niemiec, 2018; Weber et al., 2016; Linkins et al., 2015).

The construction of an ideal self arises in the stories of heroes, whether biographies of historical characters who have accomplished great deeds or intellectuals, artists, sportsmen or characters of fictional stories who constitute references to an ideal, such as fictional heroes or myths, which are symbolic representations of virtues. Stories can be powerful stimuli to the development of the ideal self in children and adolescents as well as in adults (White & Waters, 2014).

Some scientific studies have shown that identifying heroes and virtues in classical literature contributes to stimulating motivation to build the ideal self (White & Waters, 2014).

This idea of education is not original. It dates to the beginnings of Western civilization, particularly to ancient Greece (Jaeger, 1985), and unfortunately it was lost over centuries after the Middle Ages.

The *paideia* was the term used in Greece (and in early Christianity) for an educational system whose goal was the development of a citizen who sought perfection, through the development of reason as well as the development of other dimensions of learning, notably at the emotional, social and moral levels, and capable of a constructive and positive role in society (Jaeger, 1985). The Greeks distinguished the concept of reason from the concept of intelligence. In ancient Greece, reason meant the development of intelligence applied to universal concepts that involves all human beings in harmony, and not the satisfaction of personal interests and values to the detriment of others (Plato, 2007). The ideal self should seek reason in the purest Greek concept of paideia, as well as virtues, ethics, emotional balance and social experience that contribute to the greater good (Jaeger, 1985). This idea was developed by Plato (Plato, 2007).

If the ideal self did not contemplate the idea of a contribution to the greater good, it could not be considered ideal self but only selfish choices.

In the construction of the ideal self, the child and adolescent should be encouraged to broaden their vision for the whole systems around them, to the extent that they are part of and should contribute to this whole that is society in a positive and constructive way (Jaeger, 1985), not only for the conquest of an individual position with personal advantages for the individual and their family, as is now so common in Western culture. In this way, the construction of the ideal self should consider a broad view of the contribution of personal life for society and not the mere satisfaction of individualistic aspirations and selfish preferences of personal advantages in society, of social prominence in view of others (Plato, 2007).

Therefore, the paideia stresses the importance of real and symbolic stories as sources of incentives for children and adolescents for the construction of the ideal self (Plato, 2007).

In addition to the Greek tradition about the concept of education, a great Swiss educator of the 18th and 19th centuries, Johann Heinrich Pestalozzi (Pestalozzi, 1898), advocated the same idea of child development from the "inside out." (i.e., from the self to the outside social world).

In the 18th and 19th centuries in Europe, his educational method was the most popular method used by parents to educate their children. For centuries, Switzerland has been considered one of the best countries in the world to educate children, so much so that it was common for French, German and English nobles to send their children to study in Switzerland. According to this educator, the teaching method was similar to the Greek idea of "inside out," and development of the individual talents of each child was characterized by encouraging the development of intrinsic virtues and intellect, contrary to the Catholic or Protestant religious indoctrination methods that were common at the time.

This aspect was more important than cultivating the child's external knowledge or indoctrination of religious and moral values (extrinsic morals). Thus, education should be intrinsic (construction of an ideal self) and from "inside out" (i.e., from the self to the extrinsic world), and not just the indoctrination of an education system, and must contemplate an intrinsic morality and the search for inner balance (Pestalozzi, 1898). For example, in the work *How Gertrude Teaches Her Children*, Pestalozzi used the reference of a genuinely virtuous mother, characterized by an intrinsic morality, to develop in children an intrinsic aspiration to be better human beings (Pestalozzi, 1898). In addition, it stimulated autonomy in learning compared to current methods with rigid curricula not focused on students but on programmatic content, not only at the basic level but also at the secondary and higher levels.

Children were encouraged to develop individual projects of their intrinsic interest, in accordance with talents, qualities and virtues, and there was not so much emphasis on the fulfillment of pre-established technical and scientific knowledge programs (Pestalozzi, 1898).

Depending on the autonomous choices of the children, the teacher provided the knowledge needed to help them succeed in the projects of their own interest, a method similar to master's and doctoral degree programs in modern universities in the present (Pestalozzi, 1898). The difference is that he applied them to primary and secondary education.

Thus, about the "teaching method," clearly the educational methodology of Pestalozzi and ancient Greece were much more focused on the development of the child and adolescent compared to present education, whose "method" is much more focused on the "transmission" of technical and scientific knowledge than on the actual development of each student. The combination of the Pestalozzi method and ancient Greece philosophy with current technical and scientific knowledge can constitute a revolution in education in the 21st century.

Another important aspect is that several scientific studies have found that the construction of intrinsic values based on the self (values of personal development, relationships with deep meaning, and contribution to society) are much more intrinsically motivating compared to values that are extrinsic to the self, like prestige, reputation, power and extrinsic rewards (Kasser & Ryan, 1993, 1996).

The construction of the ideal self is important for the formation of children and adolescents to the extent that it constitutes a stimulus that leads to self-realization in a practical life.

And other studies have shown that if a certain value is central to the self of an individual (concept of self-centrality; Verplanken & Holland, 2002), then the consistency between the value and behavior will tend to be much higher compared with a value that is not central to the self (Verplanken & Holland, 2002).

The construction of the ideal self is a powerful stimulus for values to become central to an individual's self. When children and adolescents want to realize a central value for their sense of self through concrete behaviors on a day-to-day basis, this value becomes part of the personal identity of the child, promoting positive emotions and emotional well-being.

For example, an investigation held in Belgium found that when children internalize in their self the importance of following the rules of the classroom, they tend to possess much better behaviors compared to children who do not internalize in their self (Aelterman et al., 2019).

Other extremely relevant investigations have experimentally demonstrated that when people cultivate more intrinsic goals compared to extrinsic goals, they tend to have higher levels of well-being and happiness (Ryan et al., 1996; Deci & Ryan, 2000).

The intrinsic goals are those related to personal development, with deep meaning of relationships, including with parents, family members, friends, boyfriends and so on, and with the intention of contributing to a better society.

Extrinsic goals are related to the goal of wealth, beauty and fame, and involve mechanisms of social comparison and social approval from others, and several studies associate extrinsic values with higher levels of depression and anxiety and lower levels of well-being and happiness (Kasser & Ryan, 1993; Ryan et al., 1996).

Even in longitudinal studies, in which participants are studied over a defined period, it turns out that those who seek intrinsic goals tend to have higher levels of well-being and happiness compared to those seeking extrinsic goals (Sheldon & Elliot, 1999).

Research showed that adolescents seeking extrinsic goals of fame, money and physical image tend to engage in more behaviors that compromise health, such as tobacco use, marijuana and sexual intercourse (Williams et al., 2000), which eventually revealed the importance of intrinsic values to children and adolescents. In this way, the ideal self should be an important stimulus for the child and adolescent to establish self-relevant goals, which increases health and healthy emotions and behaviors.

Other important research has proven that students who have predominantly extrinsic goals (money, image and popularity) tended to have significantly lower self-realization, psychological vitality and happiness (Kasser & Ahuvia, 2002). And obviously, the low levels of self-realization, psychological vitality and happiness significantly affect children, adolescents and adults from an emotional and social point of view.

Research conducted in Singapore management and economics students, who usually implicitly stimulate materialistic values (Kasser & Ahuvia, 2002), showed similar results, raising the importance of business schools to intentionally stimulate intrinsic values in their training courses to prevent the negative effects of the extrinsic goals implicitly transmitted to students.

Thus, values such as personal development, contribution to the community and global society and stimulation of high-quality connections should be the school environments in which business schools involve students, to encourage young people to develop intrinsic values to promote their individual well-being and happiness, on the one hand, and their contribution to society, on the other. This is a gap in the world's leading business schools: the lack of incentive to intrinsic values to young students.

Finally, based on three experimental studies, it was concluded that the environments that stimulate the autonomy of students as well as the intrinsic values of personal development, helping society and the deep meaning of relationships tend to have a positive effect on learning, performance and persistence (Vansteenkiste et al., 2004).

Science has also revealed that the values that children and adolescents cultivate (intrinsic or extrinsic) will have an impact over the subsequent years of life after school and university, and intrinsic values increase levels of emotional well-being and happiness, and extrinsic values diminish it (Niemiec et al., 2009). Therefore, the importance of development programs of emotional and social intelligence also involves the child's parents and family, to stimulate intrinsic values (search for the ideal self) and the autonomy of students, both in the relationship between teachers and students and in the relationship between parents and children.

In one of the most significant scientific articles on motivation and education, researchers proposed that to promote motivation and increase learning outcomes, performance and school persistence, it is essential to stimulate intrinsic motivation (through genuine interests that are intrinsic to the self of each children/adolescents) and the confidence of each child in their abilities, talents and attributes (Deci et al., 1991). This modern scientific finding is perfectly in line with the idea of education in ancient Greece and the method of Pestalozzi in Switzerland mentioned earlier.

Consequently, intrinsic motivation and intrinsic values have a huge effect on emotional well-being and happiness (Kasser & Ryan, 1993, 1996; Vansteenkiste et al., 2004; Sheldon & Elliot, 1999).

In other research, it has clearly been highlighted that children stimulated to autonomy (stimulation of intrinsic motivation) in their process of learning tend to report better conceptual learning and higher levels of emotional well-being compared to children learning in environments that stimulate extrinsic motivation, i.e., in environments where teachers are more controlling (Grolnick & Ryan, 1987a). Another analysis showed that positive parenting styles in the relationship with students influence them positively in values, motivation, emotional well-being and autonomy (Grolnick & Ryan, 1989). Researchers studied three types of parenting styles:

A style that supports autonomy, on one hand, or controller style, on the other: The style that stimulates students' autonomy helps them solve problems independently and helps them choose their own interests and learning preferences. The controller style is one where parents control children's behaviors and decisions and cause children/adolescents to tend to develop more extrinsic values and extrinsic motivation to learn, seeking rewards or avoiding punishment.

A style that provides clear structure for students: In this style, parents provide clear rules, structure and a guideline of behavior and also clarify positive expectations about the child or adolescent.

A style with a good degree of involvement of parents: This may include involvement in the lives of children, or the degree to which parents are dedicated and involved in the process of the child's learning process. Previous studies have shown that parents who have a high degree of involvement in the learning process of children produce the tendency in children to want to develop intrinsic values (Grolnick & Ryan, 1987b).

Parents who adopted a parental style of autonomy stimulus tended to produce in children an autonomous self-regulation and better performance (Grolnick & Ryan, 1989). The mother's high involvement in the learning process also tended to produce better school performance in children. And a parental style that provided a clear framework for rules and behavioral guidelines tended to produce a high level of understanding of the mechanisms of social control of behavior in a positive way from the children (Grolnick & Ryan, 1989).

Second Discovery: Construction of the Real Self

The second discovery seeks to answer the following questions that children and adolescents should be encouraged to answer themselves: Who is my real self? What are my strengths and weaknesses? For reasons related to self-esteem, the human mind tends to protect itself from information that has negative information about the real self. These defense mechanisms are universal and natural (Vaillant, 1992). Sometimes these mechanisms become dysfunctional because the image that children and adolescents make of themselves does not correspond to their real self, that is, to reality.

The metaphor of boiled frog syndrome is appropriate in this case to describe reactions to the knowledge of the real self. It is said that if a person puts a frog in a bowl of boiling water, it will jump instantly, like a defense mechanism. But if an individual puts a frog in a bowl with water with normal temperature and slowly heats the water to boil, the frog will not react and will warm until death. These small adjustments to change are acceptable, but the same change made abruptly will hardly be accepted (Boyatzis, 2002).

This syndrome often occurs in children, adolescents and adults. For children and adolescents to decide to change a part of themselves, it is imperative to have a sense of what they want to keep in their real self and of what they intend to change, guided by the ideal self. Thus, it is essential that they are encouraged to increase self-awareness about these two aspects: their strengths and weaknesses, through specific points of improvement of the real self, having as reference the positive aspirations of the ideal self, so that there is a process of change and healthy psychological growth.

One of the classic mistakes in the process of personal development and psychological growth is the identification of only the negative aspects of the real self for the realization of and the underestimation of the strengths and positive aspects of the self that can be strengthened and developed, and even used to help improve the weaknesses.

Recent research has shown that when children, adolescents and adults use the positive aspects of the real self for change, this becomes much faster, easier and more efficient when compared to the focus only on the negative aspects (Peterson & Seligman, 2004; Seligman et al., 2009; Gillham et al., 2011). A positive approach in dealing with the real self is much more effective in the educational process of children and adolescents, and even adults, compared to a negative approach (Peterson & Seligman, 2004; White & Murray, 2015; White & Waters, 2014).

The educational stimulus for children and adolescents to seek knowledge of the real self should be done in the context of their surrounding environment. (How are the children and adolescents adapting to the context surrounding it? What are the aspects of the environment that the child feels facilitate their learning? And which ones make it difficult for their learning?)

It is through the analyses of discrepancies and similarities between the ideal self and the real self, based on an eminently positive approach to the relationship with children and adolescents, that it is possible to construct their personal identities consciously and healthily. And the whole construction of personal identity goes through a lifelong development plan of personal motivation, based on their intrinsic motivations, aspirations, preferences and intrinsic values of their ideal self (Boyatzis, 2002).

Third Discovery: Construction of a Self-Directed Learning Project

The third discovery is the construction of a self-directed learning project, with a personal growth mindset focus (Dweck, 2012), which tries to respond to the following question: How can I reinforce the virtues I already possess, develop the virtues that I do not own and strengthen my weaknesses? This is a powerful tool for growth mindset and allows progress to be monitored over time, failures to be identified and reflections carried out on the improvements achieved and still needed. It also allows young people to set personal development goals (intrinsic goals), which will accompany them throughout their lives.

The self-directed learning project should be completely personal, it should be noted. The children and adolescents should define the content and pace of learning, with the teachers and parents, who assume the role of mentors (Boyatzis, 2002).

Fourth Discovery: Practical Experimentation of New Behaviors

The fourth discovery constitutes the experimentation and practice of new behaviors until their consolidation (Boyatzis, 2002). At this stage, the experimentation of new behaviors should be a process of continuous improvement of children and adolescents, and the analysis of the application of virtues to real day-to-day situations should be carried out systematically and periodically. To efficiently develop new behaviors, they should seek practical situations within the reality they live in, such as school, family, friends and colleagues.

It is important that they feel they are in a safe and facilitating environment of SEL. Psychological safety creates an environment in which children and adolescents feel encouraged to experience new behaviors, perceptions, thoughts and emotions, without fear of feeling embarrassed or threatened in their self-esteem.

Fifth Discovery: Social Environments of the Self-Directed Learning Project

Children and adolescents should choose the social environments where the self-directed learning project will be put into practice. Social environments can be the school, the gym, a sports club or the very house in which they live, with family members or any other organization that will serve as a social environment and where activities from the self-directed learning project will occur (Boyatzis, 2002).

It is appropriate to choose social environments and organizations that facilitate and intrinsically motivate the child and adolescent to develop their activities in a way that is healthy and positive, without inhibitions or social constraints. It is also important to identify and avoid social environments that hinder or inhibit the practice of activities that have been planned in the self-directed learning project. These environments should therefore be a source of reinforcement and intrinsic motivation, not of obligation, embarrassment or negative effort.

To be an intrinsic source of motivation, the social environments where self-directed learning projects will be developed need to stimulate autonomy in children and adolescents. Social environments that encourage autonomy are those that respect the individual rhythm of each child and adolescent, give the right of choice over the initiatives and activities to be undertaken in their self-directed learning project and help to provide a deep meaning, surrounded by intrinsic values, throughout the whole project.

Thus, the social environments where projects are implemented should respect the perspectives of children and adolescents, express interest and affection about their feelings, aspirations and difficulties and instill self-acceptance, self-initiative and self-discipline, and not a controlled or imposed discipline.

Agents of social environments, e.g., parents, teachers, colleagues and friends, should respect the perspectives of children and adolescents, not imposing their own perspective. On the contrary, they should be open to the perspectives of each child and adolescent and demonstrate an empathic and generative listening focused on the initiative, allowing their free expression of emotions.

By decreasing the pressure on children and adolescents, the agents of social environments promote a sense of autonomy, initiative and choice on them, which encourages them to act on their own initiative, stimulating them simultaneously with intrinsic motivation and exploratory behavior, responding to the emotional information available with curiosity, interest and a less defensive attitude.

This, in turn, contributes to self-acceptance of both positive and negative experiences. This behavior of stimulating autonomy constitutes a very important social environment for the self-directed learning project to be a success. Another important aspect is the joy and affection provided by the social environment where each child and adolescent develop their self-directed learning project.

The warm relationship of affection and joy supports the need for affiliation and trust that they need to feel safe in the social environment where the self-directed learning project will be developed. And children or adolescents should be in a social environment that provides them with the resources to develop autonomously the competencies necessary to implement their projects.

Thus, the social environment must be able to meet their basic needs of autonomy, affiliation and competence, fundamental needs for the development of their intrinsic motivation (Ryan & Deci, 2000).

When social environments do not stimulate autonomy, they tend to cause external pressure and internal emotional conflicts. When they do not stimulate competence, they tend to generate low self-esteem and feelings of incompetence and inadequacy. When they do not stimulate affection, love and trust, they tend not to stimulate affiliation and provoke feelings of loneliness, abandonment and insecurity, and this generates greater difficulties in regulating emotions.

Therefore, it is important that social environments (e.g., school, family, religious organizations, sports clubs, among others) or other organizations where the self-directed learning project is developed meet the needs of autonomy, competence and affiliation to provide children and adolescents a healthy and intrinsically motivating development of their emotional and social learning (Ryan & Deci, 2000).

Sixth Discovery: Identification of Mentors Who Will Help the Child and the Adolescent to Realize Each Part of the Self-Directed Learning Project

The sixth discovery is cross-sectional and connects to all previous discoveries: the development of relationships of trust, psychological support and safety that facilitate the process of self-directed learning. According to Boyatzis (2002), the ideal is that each discovery occurs through a discontinuity, through a moment self-awareness, and also leads to a healthy sense of urgency for change and personal growth of children and adolescents (Dweck, 2012).

High-quality connections convey a sense of attachment and security for children and adolescents, and this is one of the most important human motivations for personal development and intrinsic motivations for self-directed learning (Baumeister & Leary, 1995; Ryan & Deci, 2000).

These relationships create a psychological safety social context where the progress of new behaviors is measured and assessed, and the usefulness of new learning contributes significantly to change of the real self in the direction of the ideal self. In this sense, relationships (e.g., trusted friends, parents, family, teachers) can be powerful sources of social support and facilitating factors for change and self-directed learning success (Boyatzis, 2002).

Therefore, it is important that each mentor take on a relative role of support for some part of the self-directed learning project. For example, a close trusted friend can assume the role of tutor of a physical exercise project, making regular and systematic monitoring of any sporting objective to be achieved, while the father can take on the mentoring of another self-learning project, and a teacher of your confidence can take on the tutoring of another project. So, a project can have several distinct objectives, with different mentors helping and to be followed up in the process of self-learning. The better the relationship with these guardians of your trust, the more intrinsically motivating will be the implementation process and the regular evaluation of the project's progress; consequently, the results will be improved. These relationships of trust are a positive source of reinforcement for learning (a teacher, a friend, the father, or mother, etc.), to help children and adolescents put into practice the self-directed learning plan of emotional intelligence.

Many scientific studies have demonstrated the role parents play in developing children and adolescents, but many of the results of these studies can also be applied to the behavior of teachers, friends, colleagues and mentors. For example, it is known that parents who stimulate children's autonomy in adolescents generate, consequently, a greater development of executive capacities and self-control (Bernier et al., 2010), more intrinsic motivation (Soenens & Vansteenkiste, 2005), a secure attachment style (Frodi et al., 1985), a healthier identity development (Smits et al., 2010), higher performance and a higher frequency of prosocial behaviors.

In addition, parents who strengthen autonomy in children also strengthen emotional trust of children in them. For example, research showed that when parents are controlling, their children tend not to emotionally trust their parents and not to seek emotional support in cases of necessity (Ryan & Lynch, 1989).

Therefore, parents should learn to show respect, interest and affection for children and adolescents, their feelings, aspirations and difficulties, and encourage them to self-reflection, initiate self-discipline, and not generate expectations of a social or controlling and imposed discipline (Dunsmore et al., 2013; Gottman et al., 1997). They should also demonstrate an empathic listening focused on free initiative, allowing them to freely express their emotions.

By lowering the pressure about children and adolescents, parents, or teachers, colleagues and friends will tend to promote autonomy, initiative and free choice, which instills in them a sense of autonomy and competence and, consequently, intrinsic motivation and exploratory behaviors, with greater joy, well-being and interest in self-directed learning, and a less defensive attitude and less anxiety (Cleveland & Morris, 2014; Joussemet et al., 2018).

A study carried out on experimental training intervention on parents was called "How to Talk to Children in Such a Way That They Hear and How to Listen to Children in a Way That They Speak" (Joussemet et al., 2018). This training taught parents communication techniques, such as empathic accuracy, active and generative listening, virtue-based question techniques or appreciative inquiry, to develop autonomy in children in the face of their emotional responses, and the result generated a significant increase in children's emotional autonomy and self-control (Joussemet et al., 2018).

Thus, the encouragement of parents, teachers, colleagues and friends to autonomy, competence and secure attachment of children and adolescents is a key element so that the mentoring process of the self-directed learning project is successful.

With this sixth discovery, the cycle of self-learning of emotional intelligence ends (Boyatzis, 2002). If adjustments are needed throughout the self-directed learning project, it is important to review the five discoveries again, as a cycle (Boyatzis, 2002).

Emotional Intelligence in Relationships With Others: The Relationship Between Empathy, Compassion and Benevolence

In a society with a selfish tendency, one of the ways to fight collective selfishness is to stimulate new generations to develop empathic concern, compassion and to help those in need.

These aspects are of vital importance for the mental and emotional formation of the becoming adults with emotional well-being and to be productive in the future of the world. They need to be encouraged not to think only in themselves but in the environment in which they live in and to help those in need, demonstrating empathic concern.

In this context, the competence of empathy needs to be connected to the values of benevolence and compassion, and to the values of an authentic ideal self of wanting to be a better person, to contribute positively to society. This has two positive consequences: a substantial increase in their well-being and an improvement of society, resulting from its positive and constructive social actions (Kasser, 2014).

For example, a scientific study showed that when submitted to psychological intervention in terms of value training in order to reduce extrinsic values and increase intrinsic values (growth relationships with meaning and a positive contribution to society), U.S. adolescents who had more materialistic values (fame, money and success) increased their self-esteem, compared to a control group of adolescents who held materialistic values but who were not subjected to the intervention (Kasser et al., 2014). Of course, the increase in adolescents' self-esteem has a positive effect in terms of emotional well-being (Kasser et al., 2014). One of the most widely used mechanisms in relationships based on intrinsic values is empathy.

A good suggestion to increase the practice of empathy and benevolence in children and adolescents is to encourage them to participate in volunteer work systematically and regularly. This type of work has the advantage of creating habits of benevolence and generosity in them, allowing the practice of emotional empathy, cognitive empathy – that is, it helps them to think from the perspective of others and develops genuine empathic concern for other people who are in pain. The empathic concern is a way of putting into practice the values of benevolence, compassion and intrinsic values, with practical benefits for themselves (increase in self-esteem and well-being) and external benefit for others (Snyder et al., 2016; Batson et al., 2016).

The expansion of comprehensive emotional intelligence education programs in all children and adolescents is clearly a necessity and a present challenge, in research and intervention centers, schools and governments.

This requires a refinement of the quality of scientific research of interventions and training as well as training of school staff and teachers to be able to provide them high-quality emotional intelligence education.

References

Aelterman, N., Vansteenkiste, M., & Haerens, L. (2019). Correlates of students' internalization and defiance of classroom rules: A self-determination theory perspective. *British Journal of Educational Psychology, 89*, 22–40. https://doi.org/10.1111/bjep.12213

Batson, C. D., Ahmad, N., & Stocks, E. L. (2016). Benefits and liabilities of empathy-induced altruism. *The social psychology of good and evil* (pp. 359–385). Guilford Press.

Baumeister, R. F., & Leary, M. R. (1995). The need to belong: Desire for interpersonal attachments as a fundamental human motivation. *Psychological Bulletin, 117*(3), 497. https://doi.org/10.1037/0033-2909.117.3.497

Bernier, A., Carlson, S. M., & Whipple, N. (2010). From external regulation to self-regulation: Early parenting precursors of young children's executive functioning. *Child Development, 81*, 326–339. https://doi.org/10.1111/j.1467-8624.2009.01397.x

Blasi, A. (1988). Identity and the development of the self. In *Self, ego, and identity* (pp. 226–242). Springer.

Boyatzis, R. E. (2002). Unleashing the power of self-directed learning. In R. Sims (Ed.), *Changing the way we manage change: The consultants speak.* Quorum Books.

Cleveland, E. S., & Morris, A. (2014). Autonomy support and structure enhance children's memory and motivation to reminisce: A parental training study. *Journal of Cognition and Development, 15*, 414–436. https://doi.org/10.1080/15248372.2012.742901

Cohen, J. (2001). *Caring classrooms/intelligent schools: The social emotional education of young children*. Series on Social Emotional Learning. Teachers College.

Deci, E. L., & Ryan, R. M. (1991). A motivational approach to self: Integration in personality. In R. Dienstbier (Ed.), *Nebraska symposium on motivation: Perspectives on motivation* (vol. 38, pp. 237–288). University of Nebraska Press.

Deci, E. L., & Ryan, R. M. (2000) The 'what' and 'why' of goal pursuits: Human needs and the self-determination of behavior. *Psychological Inquiry, 11*, 227–268. https://doi.org/10.1207/S15327965PLI1104_01

Dunsmore, J. C., Booker, J. A., & Ollendick, T. H. (2013). Parental emotion coaching and child emotion regulation as protective factors for children with oppositional defiant disorder. *Social Development, 22*, 444–466. https://doi.org/10.1111/j.1467-9507.2011.00652.x

Dweck, C. (2012). *Mindset: Changing the way you think to fulfil your potential.* Hachette.

Frodi, A., Bridges, L., & Grolnick, W. (1985). Correlates of mastery-related behavior: A short-term longitudinal study of infants in their second year. *Child Development, 56*, 1291–1298. https://doi.org/10.2307/1130244

Gardner, H. (2011). *Frames of mind: The theory of multiple intelligences.* Hachette.

Gillham, J., Adams-Deutsch, Z., Werner, J., Reivich, K., Coulter-Heindl, V., Linkins, M., & Seligman, M. E. P. (2011). Character strengths predict subjective well-being during adolescence. *Journal of Positive Psychology, 6*(1), 31–44. https://doi.org/10.1080/17439760.2010.536773

Goleman, D., & Senge, P. M. (2014). *The triple focus: A new approach to education.* More Than Sound.

Gottman, J. M., Katz, L. F., & Hooven, C. (1997). *Meta-emotion: How families communicate emotionally.* Lawrence Erlbaum Associates.

Grolnick, W. S., & Ryan, R. M. (1987a). Autonomy in children's learning: An experimental and individual difference investigation. *Journal of Personality and Social Psychology, 52*(5), 890. https://doi.org/10.1037/0022-3514.52.5.890

Grolnick, W. S., & Ryan, R. M. (1987b). Autonomy support in education: Creating the facilitating environment. In N. Hastings & J. Schwieso (Eds.), *New directions in educational psychology: Behaviour and motivation* (vol. 2, pp. 213–232). Falmer Press.

Grolnick, W. S., & Ryan, R. M. (1989). Parent styles associated with children's self-regulation and competence in school. *Journal of Educational Psychology, 81*(2), 143. https://doi.org/10.1037/0022-0663.81.2.143

Jaeger, W. (1985). *Early christianity and greek paideia*. Harvard University Press.

Joussemet, M., Mageau, G. A., Larose, M-P., Briand, M., & Vitaro, F. (2018). How to talk so kids will listen & listen so kids will talk: A randomized controlled trial evaluating the efficacy of the how-to parenting program on children's mental health compared to a wait-list control group. *BMC Pediatrics, 18,* 257. https://doi.org/10.1186/s12887-018-1227-3

Kasser, T. (2014). Teaching about values and goals: Applications of the circumplex model to motivation, well-being, and prosocial behavior. *Teaching of Psychology, 41*(4), 365–371. https://doi.org/10.1177/0098628314549714

Kasser, T., & Ahuvia, A. C. (2002) Materialistic values and well-being in business students. *European Journal of Social Psychology, 32,* 137–146. https://doi.org/10.1002/ejsp.85

Kasser, T., Rosenblum, K. L., Sameroff, A. J., Deci, E. L., Niemiec, C. P., Ryan, R. M., . . . & Hawks, S. (2014). Changes in materialism, changes in psychological well-being: Evidence from three longitudinal studies and an intervention experiment. *Motivation and Emotion, 38,* 1–22. https://doi.org/10.1007/s11031-013-9371-4

Kasser, T., & Ryan, R. M. (1993). A dark side of the American dream: Correlates of financial success as a central life aspiration. *Journal of Personality and Social Psychology, 65,* 410–422. https://doi.org/10.1037/0022-3514.65.2.410

Kasser, T., & Ryan, R. M. (1996). Further examining the American dream: Differential correlates of intrinsic and extrinsic goals. *Personality and Social Psychology Bulletin, 22,* 280–287. https://doi.org/10.1177/0146167296223006

Linkins, M., Niemiec, R. M., Gillham, J., & Mayerson, D. (2015). Through the lens of strength: A framework for educating the heart. *The Journal of Positive Psychology, 10*(1), 64–68. https://doi.org/10.1080/17439760.2014.888581

Loevinger, J. (1976). *Ego development*. Jossey-Bass.

Niemiec, C. P., Ryan, R. M., & Deci, E. L. (2009). The path taken: Consequences of attaining intrinsic and extrinsic aspirations in post-college life. *Journal of Research in Personality, 43*(3), 291–306. https://doi.org/10.1016/j.jrp.2008.09.001

Niemiec, R. (2018). *Character strengths interventions – A field guide for practitioners.* Hogrefe Publishing.

Pestalozzi, J. H. (1898). *How gertrude teaches her children: An attempt to help mothers to teach their own children and an account of the method.* CW Bardeen.

Peterson, C., & Seligman, M. E. (2004). *Character strengths and virtues: A handbook and classification.* Oxford University Press.

Plato (2007). *Republic*. Penguin.

Ryan, R. M., & Deci, E. L. (2000). Self-determination theory and the facilitation of intrinsic motivation, social development, and well-being. *American Psychologist, 55*(1), 68–78. https://doi.org/10.1037/0003-066x.55.1.68

Ryan, R. M., & Lynch, J. (1989). Emotional autonomy versus detachment: Revisiting the vicissitudes of adolescence and young adulthood. *Child Development, 60,* 340–356. https://doi.org/10.2307/1130981

Ryan, R. M., Sheldon, K. M., Kasser, T., & Deci, E. L. (1996). All goals are not created equal: An organismic perspective on the nature of goals and their regulation. In P. M. Gollwitzer & J. A. Bargh (Eds.), *The psychology of action: Linking cognition and motivation to behavior* (pp. 7–26). Guilford Press.

Salovey, P., & Mayer, J. D. (1990). Emotional intelligence. *Imagination, Cognition, and Personality, 9*(3), 185–211. https://doi.org/10.2190/DUGG-P24E-52WK-6CDG

Seligman, M. E. P., Ernst, R. M., Gillham, J., Reivich, K., & Linkins, M. (2009). Positive education: Positive psychology and classroom interventions. *Oxford Review of Education, 35*(3), 293–311. https://doi.org/10.1080/03054980902934563

Sheldon, K. M., & Elliot, A. J. (1999). Goal striving, need satisfaction, and longitudinal well-being: The self-concordance model. *Journal of Personality and Social Psychology, 76*(3), 482. https://doi.org/10.1037/0022-3514.76.3.482

Smits, I., Soenens, B., Vansteenkiste, M., Luyckx, K., & Goossens, L. (2010). Why do adolescents gather information or stick to parental norms? Examining autonomous and controlled motives behind adolescents' identity style. *Journal of Youth and Adolescence, 39*, 1343–1356. https://doi.org/10.1007/s10964-009-9469-x

Snyder, M., Omoto, A. M., & Dwyer, P. C. (2016). Volunteerism: Multiple perspectives on benefits and costs. In A. G. Miller (Ed.), *The social psychology of good and evil* (pp. 467–493). Guilford Publications.

Soenens, B., & Vansteenkiste, M. (2005). Antecedents and outcomes of self-determination in three life domains: The role of parents' and teachers' autonomy support. *Journal of Youth and Adolescence, 34*, 589–604. https://doi.org/10.1007/s10964-005-8948-y

Turkle, S. (1996). Who am we? We are moving from modernist calculation toward postmodernist simulation, where the self is a multiple, distributed system. *Wired Magazine, 4*(1).

Turkle, S. (2003). Technology and human vulnerability. A conversation with MIT's Sherry Turkle. *Harvard Business Review, 81*(9), 43–50.

Turkle, S. (2017). *Alone together: Why we expect more from technology and less from each other.* Hachette.

Vaillant, G. E. (1992). *Ego mechanisms of defense: A guide for clinicians and researchers.* American Psychiatric Pub.

Vansteenkiste, M., Simons, J., Lens, W., Sheldon, K. M., & Deci, E. L. (2004). Motivating learning, performance, and persistence: The synergistic effects of intrinsic goal contents and autonomy-supportive contexts. *Journal of Personality and Social Psychology, 87*(2), 246. https://doi.org/10.1037/0022-3514.87.2.246

Verplanken, B., & Holland, R. W. (2002). Motivated decision making: Effects of activation and self-centrality of values on choices and behavior. *Journal of Personality and Social Psychology, 82*(3), 434–447. https://doi.org/10.1037/0022-3514.82.3.434

Weber, M., Wagner, L., & Ruch, W. (2016). Positive feelings at school: On the relationships between students' character strengths, school-related affect, and school functioning. *Journal of Happiness Studies, 17*(1), 341–355. https://doi.org/10.1007/s10902-014-9597-1

White, M. A., & Murray, A. S. (Eds.) (2015). *Evidence-based approaches in positive education: Implementing a strategic framework for well-being in schools.* Springer.

White, M. A., & Waters, L. E. (2014). A case study of "The Good School:" Examples of use of Peterson's strengths-based approach with students. *Journal of Positive Psychology, 10*(1), 69–76. https://doi.org/10.1080/17439760.2014.920408

Williams, G. C., Cox, E. M., Hedberg, V., & Deci, E. L. (2000). Extrinsic life goals and health risk behaviors in adolescents. *Journal of Applied Social Psychology, 30*, 1756–1771. https://doi.org/10.1111/j.1559-1816.2000.tb02466.x

Chapter 9

Emotional Intelligence Exercises

Exercise 1: Self-Awareness and Emotional Self-Control in Situations

Ask the child or adolescent to reflect on the main situations that occur in their lives that cause them anger and fear. Ask them to ponder how they feel when they are influenced by these emotions. Also ask them to ponder alternatives to feeling anger and fear in these situations. Ask them to identify specific triggers that generate these emotions in these situations. Help him enumerate a set of situations that provoke similar reactions of anger and fear, so that when these situations are repeated, the child is more emotionally conscious and knows how to react with emotional intelligence. Help the child devise a self-control plan to emotionally manage these emotions generated in these situations, i.e., develop the competence of emotional self-control. Use Tables 9.1 through 9.4 as written supports.

Table 9.1 Anger-Generating Situations – Sample

Situations That Generate Anger	Situational Trigger	Physical, Verbal and Emotional Reactions	Behavioral Reactions	Emotional Self-Control Plan
Describe three concrete situations in which the emotion of anger tends to be triggered	Situation 1: When some colleague tells me some defect of mine in an aggressive or sarcastic way	I feel sweat on my hands; I get a headache; I feel a lack of energy and enthusiasm; I feel heat; I feel sad after feeling a lot of anger	I start yelling at my colleague; I do not let the colleague speak; I do not want to interact with my colleague anymore;	In the first person: I will be aware of all the situations that a colleague tells me some defect of mine aggressively, and I will reflect on the emotions I am **feeling when the emotion of anger is triggered**. I will talk mentally with the emotional state I am feeling in the sense of calming down and

(Continued)

Table 9.1 (Continued)

Situations That Generate Anger	Situational Trigger	Physical, Verbal and Emotional Reactions	Behavioral Reactions	Emotional Self-Control Plan
			I walk away from the colleague	reacting appropriately, talking to the colleague whether their comment is correct or not. If the colleague is right in their comment, humbly admit the defect and think of ways to overcome it in a practical and concrete way, or if, on the other hand, the comment is wrong and unfair, I will unaggressively verbalize the colleague's misunderstanding. Regardless of whether the colleague's intention is positive or negative, I will maintain an emotional behavior of balance and wisdom. If the colleague is malicious, this is their problem. I will maintain my emotional balance regardless of the intentions of others, whatever they may be. I will be mindful of my emotional reactions when I realize other people's negative intentions toward me. If that happens, I am going to manage my fear or my anger, or another emotional reaction I have.

Table 9.2 Anger-Generating Situations

Situations That Generate Anger	Situational Trigger	Situational Trigger	Behavioral Reactions	Emotional Self-Control Plan

Table 9.3 Table of Fear-Generating Situations

Situations That Generate Fear	Situational Trigger	Physical, Verbal and Emotional Reactions	Behavioral Reactions	Emotional Self-Control Plan

Table 9.4 Table of Fear-Generating Situations

Situations That Generate Fear	Situational Trigger	Physical, Verbal and Emotional Reactions	Behavioral Reactions	Emotional Self-Control Plan

Exercise 2: Self-Awareness and Emotional Self-Control in Relationships

Ask the child or adolescent to reflect on three of the most important relationships they have in their lives that provoke positive emotions and three of the main relationships they have in their lives that provoke negative emotions. The goal is to (1) help the child or adolescent to have greater emotional self-awareness about the elements contained in relationships that provoke positive and negative emotions and (2) learn to emotionally manage emotions to improve the quality of relationships, that is, develop the competence of emotional self-control. Use as support Tables 9.5 through 9.8. The first rows of Tables 9.5 and 9.7 provide examples.

Table 9.5 Self-Awareness of Positive Emotion–Generating Relationships – Sample

Relationships That Generate Positive Emotions	Causes Generating Positive Emotions	Physical, Verbal and Emotional Reactions	Behavioral Reactions	Emotional Self-Control Plan
Describe three concrete relationships that generate positive emotions.	I realize that this person feels genuinely happy when he's with me. I feel like I always learn something new and important to my background as a person when I'm with this person. This person makes me feel self-confident in my own capabilities.	Joy Enthusiasm Interest Curiosity Gratitude	I feel more energy when I'm with this person. I feel more like learning new things when I'm with this person. I feel more tenderness and love when I'm with this person. I feel more self-realization and joy when I'm with this person.	I will think of concrete ways to repay all the good that this person does to me, thinking about how I can help this person in things that are really important to them. I'll plan to help her, make her feel happier.

Table 9.6 Self-Awareness of Positive Emotion–Generating Relationships

Relationships That Generate Positive Emotions	Causes Generating Positive Emotions	Physical, Verbal and Emotional Reactions	Behavioral Reactions	Emotional Self-Control Plan

Table 9.7 Self-Awareness of Negative Emotion–Generating Relationships – Sample

Relationships That Generate Negative Emotions	Causes Generating Negative Emotions	Physical, Verbal and Emotional Reactions	Behavioral Reactions	Emotional Self-Control Plan
Describe three concrete relationships that generate negative emotions.	I realize that this person always tries to point out my weaknesses when he's with me. Somehow, this person always tries to show himself superior to me in the most diverse situations.	Anger Indignation Moral Disgust Sadness	I feel less energy when I'm with this person. I feel a lot of physical and emotional stress when I'm with this person. I feel aggressive when I'm with this person. I feel the desire for revenge when I am with this person, because I feel despised.	In the first person: I will be attentive to control my negative emotions whenever I am relating to this person. I will try to control my anger, my indignation, my moral disgust and my sadness when I am interacting with this person. When I feel angry, I will reflect that the emotion of anger does not help me on a personal level, nor does it help to manage the relationship intelligently with this person. I'm going to evaluate if there is any possibility to talk openly with this person and make them reflect on the pattern of behavior that person has with me, so that he changes that behavior. I will do this in a respectful, proper and intelligent way. If there is no possibility of talking openly and frankly with that person and helping in the negative pattern of relationship with this person, I will focus on managing my personal emotions about them. The others cannot determine how I'm going to feel. I am autonomous to feel positive emotions, regardless of a positive relationship or not. The positive emotions are mine, the behavior of the other is the other's problem. I have the power, autonomy and capacity to determine my own emotions.

150 Emotional Intelligence Applied to Education

Table 9.8 Self-Awareness Table of Negative Emotion–Generating Relationships

Relationships That Generate Negative Emotions	Causes Generating Negative Emotions	Physical, Verbal and Emotional Reactions	Behavioral Reactions	Emotional Self-Control Plan

Exercise 3: Impulsivity Self-Control Exercise and Delay of Gratification

Ask the child or adolescent to describe three situations in which they find it difficult to control themself in terms of impulsivity. It may be the tendency to eat too many sweets daily, the tendency to respond to something unpleasant when irritated or the difficulty in delaying a gratification, such as the difficulty in saving money on small things that prevents saving for a summer vacation. Use Tables 9.9 and 9.10 for support for the exercise.

Table 9.9 Impulsivity Self-Control Exercise – Sample

Impulsivity Situations	Situational Trigger	Physical, Verbal and Emotional Reactions	Behavioral Reactions	Emotional Self-Control Plan
Describe three concrete situations in which you find it difficult to control yourself in terms of impulsivity or delay of gratification, i.e., to delay a short-term reward in order to achieve a long-term bigger reward.	Tendency to eat more sweets daily at a specific time of day. Tendency to respond aggressively when angry. Difficulty in delaying a specific reward, such as saving on small daily things that would allow savings for an extended summer vacation.	Tendency to eat more sweets daily at a specific time of day. Tendency to respond aggressively when angry. Difficulty in delaying a specific reward, such as saving on small daily things that would allow savings for an extended summer vacation.	Start acting without reflecting. Act quickly. Speak quickly. Act before thinking and reflecting.	In the first person: For each situation described, I will establish a daily action plan with the objective of implementing the objective of controlling my impulsivity, using emotional self-awareness to note what emotions cause these impulsive behaviors and establishing concrete habits of impulsivity control.

Table 9.10 Impulsivity Self-Control Exercise

Impulsivity Situations	Situational Trigger	Physical, Verbal and Emotional Reactions	Behavioral Reactions	Emotional Self-Control Plan

Exercise 4: Exercise of Self-Discipline and Creation of Good Habits

Identify three specific habits that you would like to acquire in a practical, regular and day-to-day way that you feel you do not have and that could bring a great positive contribution to your life. Identify the emotional, personal and relational factors that hinder and facilitate the acquisition of this habit. Create an implementation plan for each habit. Use Tables 9.11 and 9.12 as support for the exercise.

Table 9.11 Exercise of Self-Discipline and Good Habits – Sample

Habits to Acquire	Personal Benefit of Habit	Benefit of Habit for Others	Implementation Plan
Identify three specific habits that you would like to acquire in a practical, regular way, in order to be applied on a daily basis.	What is the benefit of this habit for yourself?	What benefit do others around you receive?	Mondays, Wednesdays, and Fridays at X o'clock do . . .

Table 9.12 Exercise of Self-Discipline and Good Habits

Habits to Acquire	Personal Benefit of Habit	Benefit of Habit for Others	Implementation Plan

Exercise 5: Purpose of Life Exercise

Provide the following exercise to the child or adolescent. Ask the child or adolescent to work out a purpose for their personal and professional lives. To this end, consider a purpose that contains the following elements: **the development of its strengths, the correction of its points to be improved, the type of specific contribution to society**, the type of inner well-being you want to achieve and the type of concrete benefit for others. Use Tables 9.13 and 9.14 as support for the exercise. Use the information developed in Exercise 1 to draw inspiration from the strengths and points to improve.

Table 9.13 Purpose of Life Exercise – Sample

Purposes of Personal Life	Strengths	Points to Improve	Benefit to Self	Benefit to Others	Implementation Plan
Identify three specific personal life purposes you would like to achieve throughout your life.	Identify your strengths that will make it easier to accomplish these personal life purposes.	Identify your personal points that you need to improve to achieve these life purposes.	Identify what the benefits of these purposes will be for your "self" and your personal self-realization.	Identify the benefits of their life purposes for others and for society.	Establish concrete plans for the realization of these purposes of life.

Table 9.14 Purpose of Life Exercise

Purposes of Personal Life	Strengths	Points to Improve	Benefit to Self	Benefit to Others	Implementation Plan

Exercise 6: Exercise of Intrinsic Values of Life

Describe three situations in which you have helped a friend, a colleague, a partner, your parents, teachers or a stranger. Describe in detail each of the three situations. Also describe how you felt during this action. Describe how you felt after the situation occurred. Describe how the person you helped felt and describe how you felt about it. Based on these three concrete experiences, create a plan to help three people in the next 3 weeks, choose one for each week. Describe this plan in detail, considering the name of the person you intend to help, how you intend to help and what actions you will take to help each one of them. (Use Tables 9.15 and 9.16 for support.)

Situation 1:

Situation 2:

Situation 3:

Table 9.15 Exercise of Intrinsic Values of Life – Helping Others

People You Helped	Emotions During Action	Emotions of Those Who Received the Action	Emotion After Action

Person 1 you want to help (describe a plan to help person 1):

Person 2 you want to help (describe a plan to help person 2):

Person 3 you want to help (describe a plan to help person 3):

Table 9.16 Exercise of Intrinsic Values of Life – Helping Others

People You Plan to Help	How Do You Intend to Help	How Do You Intend to Help	Timing

Part III

Emotional Intelligence Applied to Organizations

Chapter 10

Leadership With Emotional and Social Intelligence

Any relatively experienced professional has lived the situation of having had an explosive and unbalanced boss from an emotional point of view. Situations of lack of emotional self-control from bosses, abuse of power and intimidation in the relationship with subordinates and with their teams are common and can be extremely disengaging for all professionals (Kellerman, 2004).

Often, the disengagement does not come from the work itself but from the emotional stress of having to deal with an authoritarian chief, who often gets emotionally out of control in many situations. And, because the chief has greater formal power, the relationship of authenticity between the chief and subordinates is compromised.

Due to fear of reprisals, the subordinate fears being punished for telling the truth, and so the subordinate pretends everything is okay most of the time. And because the boss does not receive authentic feedback from their bad leadership, there are no changes over time, and usually a bad leader with this behavioral profile tends to remain the same for many years or for a professional lifetime (Kellerman, 2004).

This means that the quality of the real relationship between leaders and subordinates tends not to evolve over time, because the leader continues to maintain their behavioral repertoires (Kellerman, 2004), and the subordinates, consequently, tend to assume a defensive mindset of "survival" and "continuity" within the organization, if they do not have a better alternative in the market.

This behavior presents several costs, particularly of workforce disengagement and the absence of an intrinsic motivation for work from the part of subordinates. These are tacit and present issues in the day-to-day life of almost all organizations.

Despite the expansion of professional leadership training around the world (revenue from leadership training programs was around US$20 billion in the United States alone, according to the American Association for Training and Development), in most organizations, problems related to behaviors of bad leaders are more frequent than positive behaviors (Kellerman, 2018; Pfeffer, 2015).

One of the ways to try to help organizations deal better with these situations is to provide them with training in leadership with emotional and social intelligence for professionals who occupy positions of formal power.

The heads who attend this type of training go through the process of a 360-degree evaluation of their leadership skills with emotional and social intelligence, i.e., they carry out assessments of themselves (self-assessments), their leaders, their peers, subordinates, and their internal and external customers. The 360-degree evaluation, when well used, is powerful to provide the chief with greater emotional self-awareness to impact their level of leadership skills with emotional and social intelligence.

Based on this report (Boyatzis et al., 2017), there are sufficient objective elements for, with the help of a mentor, a coach or trainer who is specialized in leadership and coaching with intelligence emotional and social intelligence, to help managers build a leadership personal development plan. This plan allows, in a detailed and action-oriented manner, the acquisition of leadership skills with emotional and social intelligence on the job.

The Model of Leadership With Emotional and Social Intelligence

The model of leadership with emotional and social intelligence was proposed in the book *Primal Leadership* by Daniel Goleman, Richard Boyatzis and Annie McKee (Goleman et al., 2013), composed of 18 competencies when published in 2002, and reduced to 12 in 2017 (Boyatzis et al., 2017), after a review made by Professor Richard Boyatzis and his team of researchers with the aim of increasing the predictive validity of each competence in various performance levels in organizations, such as sales volume, quality of service to the client, organizational climate, among others (Druskat et al., 2013; Boyatzis et al., 2012, 2017; O'Boyle Jr. et al., 2011).

This model is based on two major levels of analysis: focus on the self and focus on others. In the focus on self, the clusters of competencies are self-awareness and self-management. In the cluster of self-awareness, the critical competence is emotional self-awareness.

In the self-management cluster, the competencies are emotional self-control, adaptability, achievement orientation and positive outlook. In the focus on others, the clusters of competencies are social awareness and relationship management. In the cluster of social awareness, the competencies are empathy and organizational awareness. And in the cluster of relationship management, the competencies are influence, coaching/mentoring, conflict management and inspirational leadership.

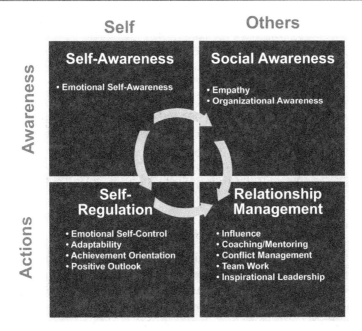

Figure 10.1 Competencies of the model of leadership with emotional and social intelligence (Goleman et al., 2013).

Each of these competencies is then defined:

Emotional self-awareness – This is the ability to identify your own emotions at the time they occur and their effect on performance.

Emotional self-control – This is the ability to control emotional impulses and negative emotional states, and to maintain efficiency in situations of stress and in hostile conditions.

Adaptability – This is flexibility in the face of change, dealing with multiple requirements and the ability to adapt their ideas and approaches to new situations.

Achievement orientation – This is orientation for results: seek to be your better self to achieve a standard of excellence and use the establishment and guidance to achieve challenging goals.

Positive outlook – This is the ability to see positive aspects in people and situations; it is using persistence to achieve objectives despite obstacles.

Empathy – This is being aware of the feelings and perspectives of others and demonstrating active attention to their interests and concerns.

Organizational awareness – This is the ability to read emotional dynamics teams, formal and formal power relations as well as dynamic implicit and explicit relationships among the actors of the organization.

Influence – This is the ability to create a positive impact on others, to persuade or convince others to have their support.

Coaching and mentoring – This is the ability to promote the learning of others through systematic feedback, monitoring, coaching and mentoring.

Conflict management – This is the ability to help others in stressful conflict situations through the management of emotions, high-quality connections, integrative/cooperative negotiation, and conflict management skills.

Teamwork – This is work with others to achieve a common goal. Create group synergies to achieve common goals.

Inspirational leadership – This is the ability to inspire people to want, by their own choice and with intrinsic motivation, to achieve the organizational goals.

Emotional Self-Awareness and Emotional Self-Control

Emotional self-awareness is the ability of human beings to become aware of the emotion felt at the very moment you are feeling it. For instance, a 360-degree report of leadership with emotional and social intelligence increases the level of self-knowledge, but while it allows the identification of strengths and weaknesses to improve leadership with emotional and social intelligence competencies in training, only emotional self-awareness enables awareness of the emotion during the emotional experience itself, in a particular situation. This is a qualitatively distinct competence from emotional self-knowledge, because although self-knowledge allows a deliberation and understanding of oneself, only emotional self-awareness allows a real change in concrete situations.

In addition, emotional self-control is impossible if there is not a preceding process of emotional self-awareness. For example, one leader can only control the emotion of fear if they have emotional self-awareness of fear when it happens in a specific situation.

In an organizational formal position, this is even more difficult, because it is common to notice that people tend to have enormous emotional self-control toward the boss (even when sometimes feeling a lot of anger toward him) but quickly lose self-control toward their subordinates, peers or persons in their personal circle (Bandura, 1978). This means that when we are attentive (emotional self-awareness) to our emotions and genuinely are willing to control them, we increase the probability of success in doing it, if we really want to (Bandura, 1978).

It is much more common for people to lose emotional self-control in the face of someone with less power compared to someone with more power. Therefore, one of the variables that increases or diminishes the desire for emotional self-control is power.

When people feel they have more power over someone, unconscious and automatically, they are less concerned about losing emotional self-control, because they know there will be no negative consequences from this loss of self-control. On the other hand, when dealing with someone with more power, people tend to be much more careful and attentive not to lose emotional self-control over this person, because they know that harm can be generated from this behavior. This is an unconscious form of moral disengagement (Bandura, 1978) toward emotional self-control, and unfortunately, it is very common in many leaders.

This means that underlying the issue of emotional self-control is the question of the relationship of power and respect for the other person in the relationship (Bandura, 1978).

An emotionally intelligent leader understands all this and maintains a relationship of great respect and consideration even for those who have less power, and builds relationships of authenticity, trust, friendship, cooperation and development with subordinates.

The relationship is seen as positive and constructive for both parties and becomes stimulating and engaging for subordinates and teams.

In addition to generating emotions of joy, well-being, admiration, gratitude and moral and intellectual elevation, it allows the construction of a dynamic, positive and motivating relationship.

The positive relationship based on authenticity fulfills the human need for affiliation and development of professional skills when it causes the development of personal and professional staff and stimulates decision-making autonomy (Meyer & Gagne, 2008; Sheldon et al., 2003).

Thus, the emotional intelligence applied to leadership means to understand that leading is an act of gradually giving power and autonomy through a relational and systematic process of subordinates' development (McGregor, 1960; Kellerman, 2007; Crossman & Crossman, 2011).

If all managers had this mindset, and not the classical orientation of power exercise through intimidation (which generates fear) of subordinates, the quality of leadership in all organizations would be far superior, and the emotions generated by both leaders as well as followers would prove to be much more positive and productive (McGregor, 1960).

Instead of generating fear, the leadership process would produce positive emotions, and the quality of the relationship between people would be deeper and more authentic, what is now called by research high-quality connections, with great productivity, and with the perspective of personal and professional development and psychological well-being (Dutton, 2003; Dutton & Heaphy, 2003; Stephens et al., 2011).

Awareness of the role of the emotionally intelligent leader in developing people in their technical and behavioral competencies constitutes a starting point for a high-performance leadership. Thus, within a sample of leaders with high technical and management skills, the differentiating factor for high performance is leadership with emotional and social intelligence, which does not mean that technical and management skills are not important.

They are crucial for the construction of high performance, but the differentiating factor for leadership processes and for the quality of relational processes is leadership with emotional and social intelligence.

Leadership with emotional and social intelligence promotes a higher frequency of positive emotions in leaders, subordinates, the organization's climate, and the quality of relationships between leader and subordinates and between the members of a team.

This does not mean that negative emotions do not have their warning function within leadership processes, but they should be used intelligently and punctually, never frequently or chronically, as the frequency of negative emotions in leadership processes is very destructive to the relationship between leader and subordinates.

High-quality relationships are those with the greatest authenticity, which cultivates positive emotions more frequently. But those in high-quality relationships also know how to deal with emotional intelligence in difficult situations, generating negative emotions with functions, intelligently and strategically.

Later in this book, in the leadership exercises' part, it will be discussed how this process can be used in a practical way in many situations. An emotionally intelligent leader tends to generate a relational dynamics with their teams that promotes a higher frequency of positive emotions such as joy, moral elevation, gratitude, admiration, interest, curiosity, among others. Scientific research shows that these positive emotions generated by leaders toward their teams promote creativity and organizational innovation (Amabile et al., 2004, 2005; Zhou & George, 2003); the improvement of the organizational climate (Boyatzis et al., 2017); the systematic development of subordinates' competencies (Amabile & Kramer, 2011); a focus on the growth mindset (Dweck, 2016); better business performance at the quantitative level, such as sales volume (Boyatzis et al., 2012); and an increase in customer satisfaction (Pugh, 2001; Kernbach & Schutte, 2005).

However, for a leader to be able to stimulate positive emotions in their teams, the leader needs to generate these emotions in themselves first and then emotionally influence their teams (Hatfield et al., 1993; Barsade, 2002).

Therefore, the model of leadership with emotional and social intelligence is based primarily on the dimensions of the self of self-awareness and self-management (emotional intelligence), only to later focus on social dimensions, such as social awareness and relationship management (social intelligence). This means that everything starts inside each of us. There is no genuine social intelligence if there is no internalized emotional intelligence before toward the self. Genuine social intelligence is nothing more than the application of emotional intelligence in the social dimension, together with the social intelligence competencies.

Therefore, it is up to the leader to cultivate a higher frequency of positive emotions, to have a refined competence of emotional self-awareness and emotional self-control, to be able to manage their own emotions and the emotions of the teams the leader works with.

To learn leadership with emotional and social intelligence is a lifetime job. The priority for the acquisition of these competencies is the increased frequency of emotional self-awareness.

In general, we are not used to focusing our attention on what emotions we are feeling every time, simply because most of us were not trained and educated since childhood to do so, and we tend to be focused on the mental content of relationships and communications and not in the emotions underlying these mental contents.

It seems a simple exercise from a cognitive point of view, but professional leadership training with emotional and social intelligence tends to leave participants surprised as to the low level of emotional self-awareness they usually have. This exercise is the starting point for the journey of acquiring emotional and social intelligence competencies. This learning process should be deep and not superficial.

Thus, to slow down the mental content and the underlying mental processes and the systematic reflection of "What emotion am I feeling right now?" or "What is the emotion I am feeling associated with this thought, situation or person(s)?" is key for those who genuinely wish to increase the competence of emotional self-awareness. This exercise requires systematic patience and genuineness to oneself. Disguising emotions in us and others will not result in an increase in emotional self-awareness, and consequently, it will not increase emotional intelligence.

In this way, the competence of emotional self-control will increase significantly if the leader significantly increases the frequency of emotional self-awareness. After all, one can only manage an emotion one was self-aware about before. For example, how can you manage the emotion of fear in a situation of uncertainty (emotional self-control) if the leader is not aware of the intensity and frequency of this fear in themselves within a situation (emotional self-awareness)?

Adaptability

There are several experimental studies that show that positive emotions tend to increase cognitive flexibility (Isen et al., 1985), creativity (Isen et al., 1987) and the repertoire of thoughts and actions (Kahn & Isen, 1993; Waugh & Fredrickson, 2006) as well as the breadth of attention (Rowe et al., 2007). These behaviors make professionals more flexible and proactive in processes and situations that require a higher level of adaptability, particularly in organizational change.

In this way, leaders who generate enthusiasm, joy, interest and curiosity in the organizational learning process (learning organization), according to Peter Senge (1995, 2014), will also be better able to promote positive emotions in subordinates and will have higher levels of adaptability of their teams toward the changes and learning processes that organizations need to achieve. A leader who, on the contrary, tends to generate fear in their teams, in a process that requires organizational change, will have much more difficulty finding cognitive flexibility, innovation and change engagement in their team. It is known that the emotion of fear usually blocks the creative processes and makes the repertoires of leadership styles more rigid, as the action generated by the emotion of fear is avoidance or inhibition (Frijda, 1987).

Take, for example, a leader who feels fear and tries to hide this emotion in front of others; because the leader possesses more power, the leader will tend transfer this emotion of fear toward their team through a manifestation of fear or anger, through authoritarian behavior or intimidation.

The first way to deal with fear before it influences the whole team is to become aware of this fear and manage it in oneself, to avoid negatively influencing others (Snodgrass, 1985; Lewis, 2000). So, when some leader systematically has anger attacks, this probably happens because the leader is having a hard time managing their own fear.

The most classic studies of neuroscience of emotions suggest that the same brain region is responsible for fear and anger, and that this region is the amygdala, with small differences in activation in other brain regions (Rosvold et al., 1954; Weiskrantz, 1956; Adams et al., 2003; Whalen et al., 2001; Fanselow & LeDoux, 1999). What should be done in situations like this is that the leader who has this tendency should receive training to help them acquire emotional self-awareness of their fear and then learn to deal with it (emotional self-control) so that fear does not transform into anger and, consequently, does not deteriorate the quality of the relationship with the team.

Fear-based leadership is one of the most toxic leadership processes, and it paralyzes the whole organization in its capacity for adaptability at the individual, team and organizational levels.

Typically, an organization struggling to adapt to new market demands usually has leaders with limited adaptability competencies, and this negatively influences the entire organization.

Douglas McGregor (1960) even stated that an organization's limit is the limit of the brain connections of your boss. If the boss is limited and has a lot of formal power, the entire organization may be negatively compromised. In this case, there are only two outputs: providing leadership training with emotional and social intelligence for these leaders, addressing these situations, or inviting these professionals with this behavioral profile to leave the organization.

Unfortunately, we already had the opportunity to hear from several bosses in various organizations in Portuguese-speaking countries, notably in Portugal, Brazil and Angola, who said that they "like" their subordinates to be afraid of them. It gives them a feeling of power and control.

What these leaders do not know is that "power (exercised) by fear" is extrinsic (external) only from a formal power conferred by the organization and not an intrinsic power generated by a constructive emotional leadership relationship composed by the act of giving stimulus and autonomy to the healthy development of the followers (Ryan & Oestreich, 1991; Meyer & Gagne, 2008; Sheldon et al., 2003).

The leader who leads by fear holds the belief that giving power and autonomy will cause them to lose power, but the emotionally intelligent leader knows that if you give power and autonomy to the subordinate according to their motivation, intention and performance, the leader will gain the informal power of inspirational and positive leadership in a natural and non-manipulative way (Fernandez & Moldogaziev, 2015).

In the first case, the leader who leads by fear is afraid of losing power, but in the second, the emotionally intelligent leader feels joy and enthusiasm in giving power and in developing people, depending on their abilities, and performance, building positive and constructive relationships with their subordinates. These relationships generate a positive virtuous circle of high performance, identified in research as high-quality connections (Dutton, 2003; Dutton & Heaphy, 2003; Stephens et al., 2011) and spirals of positive and high-performance emotions (Sekerka & Fredrickson, 2008).

Thus, the emotionally intelligent leader seeks to systematically develop the technical and behavioral competencies of their subordinates, and depending on their performance, it creates a systematic development relationship whose objective is to gradually give more power and autonomy to the follower, which generates a high level of intrinsic motivation (Fernandez & Moldogaziev, 2015).

The growth mindset in the leadership relationship (Dweck, 2016), which is a vision of long-term personal and professional development through the leadership relational process, will build a team, within 1 year, for example, to develop a lot, compared to common leaders only interested in the subordinate's achievement of the organizational goals of the company.

In this way, we can enumerate key questions to measure the level of quality of an emotionally intelligent leader:

How much was the leader able to develop the team from a technical point of view within 1 year, and how much power and autonomy have subordinates been granted, depending on their performance?
Within 1 year, have the subordinates learned substantially new competencies or are they the same compared to last year?
Are subordinates who have been working with this leader for 1 year energized and have the vitality to innovate and learn, or are they just routinely trying to manage the work?

Achievement Orientation

The achievement orientation competence is the ability to guide behavior toward efficiency and results, showing high levels of performance and excellence. This behavior is focused on results, and the leaders who have it tend to consistently achieve high levels of quantitative and qualitative performance and results. This competence was studied scientifically by Professor David McClelland (McClelland, 1961).

The high achievement orientation leaders tend to achieve highly challenging goals and high standards of performance and tend to be more entrepreneurial (McClelland, 1965), and they tend to take responsibility for the high performance and risk-taking in a manner that is entrepreneurial and innovative (McClelland, 1965).

Positive Outlook

One of the most notable researchers in the field of learned optimism is Martin Seligman from the University of Pennsylvania (Seligman, 2006). Learned optimism is the ability to learn how to respond to adversities in a positive way in which the cause of the situation is explained and has three features (Seligman, 2006):

Permanence – Optimistic people believe that negative events are temporary and non-permanent, and recover rapidly from failure, while pessimists tend to take longer to do so.

Generalization – Optimistic people tend to compartmentalize failures (pessimists tend to generalize a failure in an area to all areas of the individual) and expand the positive aspects in an area of their life to all other areas (pessimists do the reverse).

Personalization – Optimists attribute the causes of a negative event to situational factors as the pessimists take responsibility for themselves. Optimists therefore show a propensity to possess more self-confidence. They also end up internalizing positive events while the pessimists outsource them.

In the context of leadership with emotional and social intelligence, the optimist pursues a goal in a positive and constructive manner despite all obstacles. The optimist also has the ability to help develop strengths and positives in subordinates. An optimistic leader is one who systematically seeks the best in their subordinates, promoting the intrinsic motivation in the team. Scientific studies have demonstrated the positive impact of optimism on performance of leaders and commercial teams (Schulman, 1999).

The willingness to be optimistic, i.e., to predict future positive results of events, has been related to activation of the brain regions of the anterior cingulate cortex and amygdala, compared to pessimism (Sharot et al., 2007). In addition to leadership with emotional and social intelligence, optimism helped to give rise to the positive leadership model (Brun et al., 2016) and is considered one of the virtues studied on a scientific basis (Peterson & Seligman, 2004).

Empathy

Empathy, from a scientific point of view, can be divided into three fundamental components (Zaki & Ochsner, 2012):

Cognitive empathy – This is the ability to mentally infer the situation and perspective of the other or to mentally infer the emotional state of the other.
Emotional empathy – This is the ability to emotionally feel what the other is feeling. It is the emotional sharing of experience. In a famous neuroscientific study conducted by Tania Singer, a professor of neuroscience at Max Planck Institute in Berlin and Leipzig, Germany, when subjects were given a signal that their loved ones, present in the same room, received a pain stimuli, the same brain regions were activated compared to when they themselves received a pain stimuli, such as the bilateral anterior insula, the anterior cingulate cortex, the brainstem and the cerebellum, demonstrating the phenomenon of emotional and neural empathy (Singer et al., 2004).
Empathic concern – This is the motivation to help the people with whom we feel empathy, the tendency toward prosocial behavior after we feel empathy. Therefore, empathic concern is the embryo of altruism and compassion from an evolutionary point of view (Zaki & Ochsner, 2012). A leader who has empathic concern is one who cares about the state of their followers, and the leader looks to help them in their development and their personal and emotional needs. A good leader is always the one who stimulates a growth mindset (Dweck, 2016) and helps their followers in their daily progress, promoting an intrinsic motivation, creativity and organizational innovation (Amabile & Kramer, 2011) and demonstrates through a relationship a genuine empathic concern toward its followers (Zaki & Ochsner, 2012). The area of scientific research about compassion in leadership has been greatly developed in recent years (Worline & Dutton, 2017).

Organizational Awareness

Organizational awareness is the ability to read the emotional and behavioral dynamics of teams, the formal and informal power relations and power network within the organization, as well as the implicit and explicit dynamics of actors within organizations.

An emotionally intelligent leader is usually aware of the formal and informal structures of power of the organization, where it operates, as well as the formal and informal decision-making processes, in order to learn how to position themself adequately in the face of the most varied situations and people within the organization.

The higher the formal position of a leader within an organization, the greater the need for organizational awareness competence for the leader to be effective at moving in the political arena of the organization (Pfeffer, 1992, 2010), at the ascending level, but also at the lateral and descending levels.

Organizational awareness is related not only to understanding the formal network of informal power relations within the organization but also to dealing with and using power in the most productive, positive and appropriate way.

Jeffrey Pfeffer, professor of organizational behavior at Stanford University, is one of the most important experts in this competence of organizational awareness, together with the issues of power and influence in the field of leadership in organizations (Pfeffer, 1992, 2010).

Influence

Influence, according to the research of Professor Richard Boyatzis, is the most important competence of the relationship management cluster of the leadership model with emotional and social intelligence (Boyatzis et al., 2017).

Influence is the ability to have a positive impact on others, to persuade or convince them and to have their support. It has a whole area of scientific research dedicated to it, with hundreds of experimental studies.

Social influence can occur automatically and unconsciously in any social interaction. When social influence is made in an intentional and strategic way by the influencer, it is called persuasion, and it can be applied to leadership, marketing and sales, and in close relationships or any social interaction that requires persuasion (Cialdini, 1984).

The scientific concept of persuasion covers the ability to induce others to do what we want, in such a way that they feel that the decision was made by them without feeling pressured to do so (Cialdini, 1984). Fundamentally, there are six major strategies of influence: social proof, liking, authority, reciprocity, scarcity/rarity and consistency (Cialdini, 1984).

Strategies of Influence: Social Proof

People tend to be persuaded to follow other people's behavior when other people have previously carried out this behavior. If someone is important, i.e., a social reference in each society, and carries out a behavior, people of this society will tend to imitate this person. For example, if a movie artist with high social reputation has sustainable consumption habits, it will tend to convince many people to adhere to sustainable consumption habits as well. This is, therefore, the concept of social proof.

Strategies of Influence: Liking

The liking strategy of influence involves the use of strategies to please those you want to convince. Research has shown that people tend to more easily be convinced by those who like us compared to those who do not like us. There are four major liking persuasion strategies: physical attraction, perception of similarity, the effect of familiarity (prior contact, personal relationship, etc.) and compliments (Cialdini, 1984).

Strategies of Influence: Authority

The influence strategy using authority is based on the idea that the probability of being persuaded by someone we consider an authority is much bigger compared to an ordinary person. There are three main types of authority: technical, intellectual or competence authority and formal authority or symbolic authority.

Strategies of Influence: Reciprocity

This strategy consists of the tendency to reciprocate whenever someone helps us or gives us something. It is an important social rule, present in all human societies. In this way, we are much more likely to be persuasive toward someone we helped before, because this person will automatically have the tendency to reciprocate the favor or help the way we did for them. If a leader shows genuine interest in helping and developing a team member, the probability of this leader to persuade the team member will be much higher, compared to a leader who does not show empathic concern or helping behaviors toward this team member.

Strategies of Influence: Scarcity/Rarity

It is the tendency of human beings to value one more product or something that is scarcer or rarer. Thus, if someone wishes to be persuasive, they should try to assign more of these attributes of scarcity or rarity to increase interest and the perceived value of a product, with the aim of increasing its potential for influence and persuasion. There are two types of rarity: the quantity limit for the acquisition of product or service and the time limit for its purchase.

An emotionally intelligent leader properly uses influence strategies to gain support for their leadership initiatives and to have a greater and better impact on the organization through attributing scarcity and rarity in leadership processes with their team. For example, attributing a sense of urgency for an excellent job is a scarcity strategy of persuasion toward the team. Another example is when a leader offers a rare opportunity for growth to a team member, to convince the team member to accept a new job role. It must be genuine and ethically based, otherwise it can generate the opposite effect on the team members.

Strategies of Influence: Consistency

The consistency-based influence strategy assumes that when people act or think in a certain way, there is a tendency to be consistent with this way of thinking or acting. For example, when a commercial salesman can get client agreement on the specific advantages of a particular product, this will tend to increase the likelihood that the customer will buy the product, because the client will not want to appear inconsistent with their initial agreement. In the leadership domain, when trying to convince others of the leader's ideas, gathering agreement on many topics before presenting the whole idea will stimulate others to want to appear consistent with previous agreements, increasing the probability of being persuasive in your message.

Coaching/Mentoring

Coaching and mentoring techniques can be used to help followers in their development of technical and behavioral competencies. A leader with strong coaching and mentoring skills is one who evaluates, plans and organizes the systematic development of the technical and behavioral skills of their followers and accompanies them regularly in their development, giving them systematic feedback and adhering to their leadership style according to the level of maturity of their followers' technical and behavioral skills.

The process of coaching and mentoring followers goes through six stages. The first is the objective assessment of the maturity level of each follower through the measurement of the level of technical and behavioral competencies. The second is the identification of the leadership style with emotional intelligence best suited to each follower, according to maturity level. The third stage is the elaboration of a coaching and mentoring plan for each follower, according to their developmental needs of competencies, from both technical and behavioral points of view.

The fourth stage is the start-up interview of the follower coaching and mentoring process, adjusted and negotiated with the follower, according to their inputs and suggestions for their development. Here, the higher the maturity of the follower, the greater is the autonomy and influence of the follower in the construction of the plan.

The fifth step is the monitoring on the ground of the previously drafted plan and systematic performance feedback of technical and behavioral performance. The sixth and final step is evaluation of the implementation of the developmental plan.

Briefly, these are the stages of the coaching and mentoring process:

1. Assess the follower's technical and behavioral competencies.
2. Identify the most appropriate leadership style suitable to the maturity level of the follower.
3. Prepare for the coaching and mentoring plan with the follower.
4. Conduct the start-up interview with the follower to discuss and build the coaching and mentoring plan, and depending on the follower's maturity, they can negotiate the building of the plan.
5. Follow-up with the follower on the ground and review systematic feedback of performance.
6. Evaluate the implementation of the developmental plan after its completion.

Scientific research shows that coaching and mentoring processes most promote the intrinsic motivation of followers, because followers feel that they are learning new technical and behavioral skills and progressing daily (Amabile & Kramer, 2011).

There is a whole area of research and application of the coaching/mentoring process called high-quality connections, developed by Professor Jane Dutton, from the University of Michigan. According to Dutton, high-quality connections are high-quality relationships that are established in organizations and promote high performance and high levels of motivation and job satisfaction (Roberts et al. al., 2005; Stephens et al., 2011; Brueller & Carmeli, 2011).

Thus, leadership must be a process that promotes high-quality connections between the leader and the follower, and the coaching/mentoring process represents a fundamental activity for that to happen.

Conflict Management

Conflict management is the ability to effectively manage conflicts and help others in situations of tension, conflict and stress, through the management of emotions, relationships, negotiation, persuasion and conflict management strategies. To understand conflict management strategies in detail, see the classic model proposed by Kenneth Thomas (Thomas, 1976; Thomas & Kilmann, 1974).

This model explains in detail the strategies of competition, avoidance, accommodation, commitment and collaboration. However, this book does not go into detail about each of these conflict management strategies, because the focus is on emotional intelligence competencies applied to conflict management.

A scientific study found that nursing professionals with high levels of emotional intelligence tend to prefer the conflict resolution strategy of collaboration (Morrison, 2008).

The essential competencies for this are emotional self-awareness, emotional self-control, adaptability to situations and people, positive outlook (optimism), taking the perspectives of all parties involved (cognitive empathy), organizational awareness and influence.

In this way, emotional intelligence competencies are crucial to managing a conflict effectively. The first questions that the leader should ask themselves in a conflict situation are as follows:

- What emotion does this conflict situation cause me?
- Does the emotion I feel help or hinder the process of managing this conflict?
- What emotions do I feel about the people involved?
- How can I control these emotions the moment they occur?
- What emotions do I need to create to better manage this conflict?

What is the most appropriate strategy to manage this conflict effectively? To this end, the competence of emotional self-control is crucial for any conflict situation.

Teamwork

Competence in teamwork is the ability to work effectively with others in a common and interdependent goal, to participate actively and constructively in team activities and to share responsibilities and rewards contributing to improvement of the achievement skills and momentum of the team. Teamwork competence has two major dimensions: the ability to lead the team effectively and the ability to make the team develop on its own, i.e., autonomously, with the aim of achieving greater technical and behavioral maturity and, therefore, greater autonomy; this means promoting autonomous or self-directed teams (Druskat & Wheeler, 2004).

The area of scientific research of high-performance teams has reached significant importance in recent years. A group of researchers created a specific model of emotional intelligence of teams (Druskat & Wolff, 2001; Wolff et al., 2002).

However, we do not detail this model because of its complexity and depth. The reader is advised to consult the *Harvard Business Review* article of Professor Vanessa Druskat for more details on this model of emotional intelligence of teams (Druskat & Wolff, 2001; Wolff et al., 2002).

Inspirational Leadership

Inspirational leadership is the ability to inspire and lead people and teams to achieve objectives, to experience the intrinsic values of personal development, relationships with deep meaning and help to others, as well as shared goals toward a meaningful mission for the organization.

An inspiring leader speaks of intrinsic values and experiences them daily in the organization and in relationships, inspiring subordinates to follow their behavior and values, because they lead by example. Being a living example for the team stimulates emotions of moral elevation, admiration, gratitude, curiosity and interest in followers.

The behaviors of a leader like Nelson Mandela can stimulate the emotion of moral elevation, and those of Steve Jobs can stimulate the emotions of admiration for talent, curiosity and interest as well as the intrinsic motivation for innovation. The most important aspect to develop this competence is the ability to be consistent between the values the leader reports and their observable behavior on a day-to-day organizational life. A living example is the only way to inspire others. Discourses without consistent behaviors can cause the opposite effect of moral disengagement and cynicism.

Styles of Leadership With Emotional and Social Intelligence

In an article published in the *Harvard Business Review*, Daniel Goleman identified six leadership styles with emotional and social intelligence, under the assumption that the emotionally intelligent leader must be flexible enough to apply each leadership style according to the demands of the situation and according to the level of maturity of each follower, in order to be effective in responding to the complex situations of everyday challenges and maturity idiosyncrasies of followers with whom the leader is confronted daily (Goleman, 2000).

According to Daniel Goleman, within a sample of leaders with the same level of technical and management competencies, which distinguishes the high-performance leader from the medium- or low-performance leader are the competencies of leadership with emotional and social intelligence. The combination of 12 competencies of leadership with emotional and social intelligence constitutes a repertoire as a set of resources that allow the leader a higher level of adaptability to complex situations and flexibility toward different levels of performance of followers, with the aim of being effective in achieving the organizational goals and developing the team effectively (Goleman, 2000).

For example, there are situations that require high levels of empathy, particularly the styles of leadership whose objective is the long-term development of followers, different from styles whose goal is to achieve short-term organizational results, in which many times the excess of empathy may be inappropriate. The six leadership styles with emotional intelligence proposed by Daniel Goleman (2000) are presented here.

Coercive Style

This style is best suited for crisis situations, where urgency is required in the decision-making processes, and in which followers must obey, because there is no time for the organization to reflect together. The main purpose of the coercive style is to obtain immediate obedience from the followers and the organization is in a situation that requires an urgent response.

When the organization is at risk of bankruptcy and needs a leader who makes fast decisions, the coercive style is the most appropriate. Of course, the use of the coercive style should never extend to the long term because the impact on followers and in the organizational climate can be very negative. The overuse of this style can be disastrous for the process for the satisfaction of followers and for the organization. The role of the leader in the coercive style is to demand unconditional and immediate obedience from the employees when the survival of the organization is at risk, or the levels of performance are too low.

The emotional intelligence competencies underlying the coercive style are emotional self-control, achievement orientation and the strategic decrease of empathy. Increasing levels of empathy competence make it difficult for the leader, in an urgent situation, to assume the coercive style, since the focus should be on the task, to assure the organization's survival, not on the people. Here, the keyword is flexibility.

Of course, this style has a negative impact on the organizational climate because it generates a lot of psychological and emotional pressure on followers. If this style is applied over a long period of time, it can generate many emotional, motivational and relational burnout in the organization. Nevertheless, the coercive style is appropriate when the situation requires rapid and urgent changes in the organization and when the performance of followers and the organization is too low.

Another situation in which the coercive style is appropriate is when followers have a low level of accountability, and they need to feel the emotional effects of the consequences of their actions and performance to be opened and aware to the need for urgent change. Thus, the follower needs an "emotional shock" to awaken their self-awareness to the urgency of profound behavioral changes and to their ability to take responsibility for their actions. Accountability is an essential competence in organizations, and the coercive style, when applied together with the visionary style, allows the leader to change the mental model, emotional vision and attitude of followers in the face of their actions. In this way, the focus of the coercive style should be on the task and not on the relationship with followers.

Visionary or Authoritative Style

In the visionary or authoritative style, the leader creates and mobilizes people for a common vision and articulates the course to be followed by the team, leaving followers free to choose how they will achieve the goals generated by this vision. So, this style helps to keep subordinates more intrinsically motivated to carry out a project.

This leadership style is most effective when the organization needs a change of vision for its future. This style is one of the most impactful on the organization. Leaders who adopt a visionary or authoritative style have to a high degree the following competencies of leadership with emotional and social intelligence: emotional self-awareness, empathy, conflict management and inspirational leadership.

Affiliative Style

This leadership style has the main objective of consolidating relations between people and building cohesion and trust within the team. It creates harmony and establishes bonds and trust among its members. The leader who uses this style values people and their feelings, and for this type of leader, it is important how they relate to each other.

It should be noted that a good organizational climate, harmony and cohesion do not necessarily have a direct impact on performance, but this is a very important aspect of organizational life, as the cohesion of the team is a necessary condition, but not sufficient, for the productivity and performance of followers.

The main competencies of the affiliative style are emotional self-awareness, conflict management, teamwork and inspirational leadership. The impact of this leadership style in the organizational climate is very positive.

Democratic Style

The leader who uses the democratic style aims to seek consensus among everyone and stimulate team participation in the decision-making processes.

It is used in decisions when team consensus is important, when the leader is not sure about the direction to be taken by the organization and needs innovative ideas from competent employees, and when the information is critical to the organizational adjustment to the processes of change implementation from a strategic plan. The use of this style of leadership has a very positive impact on the organizational climate.

The competencies of the democratic style are emotional self-awareness, empathy, teamwork, conflict management and inspirational leadership.

Pacesetting Style

The main feature of this style is the establishment of high-performance standards and should only be applied to achieve high-performing, quick results in highly motivated and competent teams.

If these two conditions are not met, this style could be disastrous, generating disengagement, burnout and negative organizational climate between the team members and the leader.

In addition, this style should not be always used. It should only be considered in specific moments and situations, or with a high-performing team or individuals.

The main phrase of a leader who adopts this style is, "Do what I do, now." The main competencies of the pacesetting style are achievement orientation as well as a strategic decrease in empathy. It should only be used in the long term in followers with high levels of performance or high potentials. In medium or low performers, this style tends to cause demotivation and burnout because followers with this level of maturity cannot handle the pressure, so it is not advisable or should be applied only temporarily in short-term projects.

For example, for a coach who is training a high-performing athlete for the Olympic Games, this style is the most suitable, because the athlete, who has high levels of performance, supports the pressure.

Toward medium- or low-performance athletes, this leadership style is not the most appropriate. Similarly, in organizations, this style is suitable only for high-performance professionals who can stand the pressure, otherwise it can cause negative emotional effects and burnout.

Coaching Style

The leader who adopts this style has a long-term view for the development of followers. Its main objective is to develop people for the future, to help them improve their performance, competencies, and talents in the long term. Thus, the leader helps employees identify their individual strengths and tries to help followers achieve their personal aspirations through the development of their real performances. The impact of this leadership style is very high; however, it takes time for effective implementation. The main competencies of this style are emotional self-awareness, empathy, coaching and mentoring, inspirational leadership, conflict management and influence.

Table 10.1 Leadership Styles With Emotional and Social Intelligence

Leadership Styles	Style Goal	Competencies of Each Style
Coercive	Demand immediate obedience in the face of an emergency	Emotional self-control Achievement orientation Strategic decrease in empathy
Visionary or authoritative	Create and convey a vision	Emotional self-awareness Empathy Influence Conflict management Inspirational leadership
Affiliative	Create cohesion and trust in the team and organization	Emotional self-awareness Empathy Conflict management Teamwork Inspirational leadership
Democratic	Creates involvement and participation of leaders in decision-making processes	Emotional self-awareness Empathy Conflict management Teamwork Inspirational leadership
Pacesetting	Establish a strong and fast pace of high performance	Emotional self-control Achievement orientation Strategic decrease in empathy
Mentor/coach	Develop at individual level people with depth at the technical and behavioral levels	Emotional self-awareness Empathy Coaching and mentoring Conflict management Influence Inspirational leadership

References

Adams, R. B., Gordon, H. L., Baird, A. A., Ambady, N., & Kleck, R. E. (2003). Effects of gaze on amygdala sensitivity to anger and fear faces. *Science, 300*(5625), 1536–1536. https://doi.org/10.1126/science.1082244

Amabile, T. M., Barsade, S. G., Mueller, J. S., & Staw, B. M. (2005). Affect and creativity at work. *Administrative Science Quarterly, 50*(3), 367–403. https://doi.org/10.2189/asqu.2005.50.3.367

Amabile, T., & Kramer, S. (2011). *The progress principle: Using small wins to ignite joy, engagement, and creativity at work.* Harvard Business Review Press.

Amabile, T. M., Schatzel, E. A., Moneta, G. B., & Kramer, S. J. (2004). Leader behaviors and the work environment for creativity: Perceived leader support. *The Leadership Quarterly, 15*(1), 5–32. https://doi.org/10.1016/j.leaqua.2003.12.003

Bandura, A. (1978). The self-system in reciprocal determinism. *American Psychologist, 33*(4), 344. https://doi.org/10.1037/0003-066x.33.4.344

Barsade, S. G. (2002). The ripple effect: Emotional contagion and its influence on group behavior. *Administrative Science Quarterly, 47*(4), 644–675. https://doi.org/10.2307/3094912

Boyatzis, R., Guise, S., Hezlett, Kerr, P., & Lams, S. (2017). *Emotional and social competency inventory research guide and technical manual.* Korn Ferry Technical Manuals.

Boyatzis, R. E., Good, D., & Massa, R. (2012). Emotional, social, and cognitive intelligence and personality as predictors of sales leadership performance. *Journal of Leadership & Organizational Studies, 19*(2), 191–201. https://doi.org/10.1177/1548051811435793

Brueller, D., & Carmeli, A. (2011). Linking capacities of high-quality relationships to team learning and performance in service organizations. *Human Resource Management, 50*(4), 455–477. https://doi.org/10.1002/hrm.20435

Brun, P. H., Cooperrider, D., & Ejsing, M. (2016). *Strengths-based leadership handbook.* Crown Custom Publishing, Inc.

Cialdini, R. B. (1984). *Influence: The psychology of persuasion.* Harpercollins.

Crossman, B., & Crossman, J. (2011). Conceptualising followership – a review of the literature. *Leadership, 7*(4), 481–497. https://doi.org/10.1177/1742715011416891

Druskat, V. U., Mount, G., & Sala, F. (2013). *Linking emotional intelligence and performance at work: Current research evidence with individuals and groups.* Psychology Press.

Druskat, V. U., & Wheeler, J. V. (2004). How to lead a self-managing team. *MIT Sloan Management Review, 45*(4), 65. https://doi.org/10.1109/emr.2004.25133

Druskat, V. U., & Wolff, S. B. (2001). Building the emotional intelligence of groups. *Harvard Business Review, 79*(3), 80–91.

Dutton, J. E. (2003). *Energize your workplace: How to create and sustain high-quality connections at work* (vol. 50). John Wiley & Sons.

Dutton, J. E., & Heaphy, E. D. (2003). The power of high-quality connections. In K. S. Cameron & J. Dutton (Eds.), *Positive organizational scholarship: Foundations of a new discipline* (vol. 3, pp. 263–278). Berrett-Koehler Publishers.

Dweck, C. (2016). What having a "growth mindset" actually means. *Harvard Business Review, 13*, 213–226.

Fanselow, M. S., & LeDoux, J. E. (1999). Why we think plasticity underlying Pavlovian fear conditioning occurs in the basolateral amygdala. *Neuron, 23*(2), 229–232. https://doi.org/10.1016/s0896-6273(00)80775-8

Fernandez, S., & Moldogaziev, T. (2015). Employee empowerment and job satisfaction in the US Federal Bureaucracy: A self-determination theory perspective. *The American Review of Public Administration, 45*(4), 375–401. https://doi.org/10.1177/0275074013507478

Frijda, N. H. (1987). Emotion, cognitive structure, and action tendency. *Cognition and Emotion, 1*(2), 115–143. https://doi.org/10.1080/02699938708408043

Goleman, D. (2000). Leadership that gets results. *Harvard Business Review, 78*(2), 4–17.

Goleman, D., Boyatzis, R. E., & McKee, A. (2013). *Primal leadership: Unleashing the power of emotional intelligence*. Harvard Business Review Press.

Hatfield, E., Cacioppo, J. T., & Rapson, R. L. (1993). Emotional contagion. *Current Directions in Psychological Science, 2*(3), 96–100. https://doi.org/10.1111/1467-8721.ep10770953; https://doi.org/10.1002/(SICI)1099-1379(200003)21:2<221::AID-JOB36>3.0.CO;2-0

Isen, A. M., Daubman, K. A., & Nowicki, G. P. (1987). Positive affect facilitates creative problem solving. *Journal of Personality and Social Psychology, 52*, 1122–1131. https://doi.org/10.1037/0022-3514.52.6.1122

Isen, A. M., Johnson, M. M., Mertz, E., & Robinson, G. F. (1985). The influence of positive affect on the unusualness of word associations. *Journal of Personality and Social Psychology, 48*(6), 1413. https://doi.org/10.1037/0022-3514.48.6.1413

Kahn, B. E., & Isen, A. M. (1993). The influence of positive affect on variety seeking among safe, enjoyable products. *Journal of Consumer Research, 20*(2), 257–270. https://doi.org/10.1086/209347

Kellerman, B. (2004). *Bad leadership: What it is, how it happens, why it matters*. Harvard Business Review Press.

Kellerman, B. (2007). What every leader needs to know about followers. *Harvard Business Review, 85*(12), 84.

Kellerman, B. (2018). *Professionalizing leadership*. Oxford University Press.

Kernbach, S., & Schutte, N. S. (2005). The impact of service provider emotional intelligence on customer satisfaction. *Journal of Services Marketing, 19*(7), 438–444. https://doi.org/10.1108/08876040510625945

Lewis, K. M. (2000). When leaders display emotion: How followers respond to negative emotional expression of male and female leaders. *Journal of Organizational Behavior, 21*(2), 221–234.

McClelland, D. C. (1961). *The achieving society*. D. Van Nostrand Company Inc.

McClelland, D. C. (1965). N achievement and entrepreneurship: A longitudinal study. *Journal of Personality and Social Psychology, 1*(4), 389. https://doi.org/10.1037/h0021956

McGregor, D. (1960). *The human side of enterprise*. McGraw-Hill.

Meyer, J. P., & Gagne, M. (2008). Employee engagement from a self-determination theory perspective. *Industrial and Organizational Psychology, 1*(1), 60–62. https://doi.org/10.1111/j.1754-9434.2007.00010.x

Morrison, J. (2008). The relationship between emotional intelligence competencies and preferred conflict-handling styles. *Journal of Nursing Management, 16*(8), 974–983. https://doi.org/10.1111/j.1365-2834.2008.00876.x

O'Boyle, E. H., Jr., Humphrey, R. H., Pollack, J. M., Hawver, T. H., & Story, P. A. (2011). The relation between emotional intelligence and job performance: A meta-analysis. *Journal of Organizational Behavior, 32*(5), 788–818. https://doi.org/10.1002/job.714

Peterson, C., & Seligman, M. E. (2004). *Character strengths and virtues: A handbook and classification*. Oxford University Press.

Pfeffer, J. (1992). *Managing with power: Politics and influence in organizations*. Harvard Business School Press.

Pfeffer, J. (2010). *Power: Why some people have it and others don't*. Harpercollins.

Pfeffer, J. (2015). *Leadership BS. Fixing workplaces and careers one truth at a time*. Harpercollins Publishers.

Pugh, S. D. (2001). Service with a smile: Emotional contagion in the service encounter. *Academy of Management Journal, 44*(5), 1018–1027. https://doi.org/10.5465/3069445

Roberts, L. M., Spreitzer, G., Dutton, J., Quinn, R., Heaphy, E., & Barker, B. (2005). How to play to your strengths. *Harvard Business Review, 83*(1), 74–80.

Rosvold, H. E., Mirsky, A. F., & Pribram, K. H. (1954). Influence of amygdalectomy on social behavior in monkeys. *Journal of Comparative and Physiological Psychology, 47*(3), 173. https://doi.org/10.1037/h0058870

Rowe, G., Hirsh, J. B., & Anderson, A. K. (2007). Positive affect increases the breadth of attentional selection. *Proceedings of the National Academy of Sciences, 104*(1), 383–388. https://doi.org/10.1073/pnas.0605198104

Ryan, K. D., & Oestreich, D. K. (1991). *Driving fear out of the workplace: How to overcome the invisible barriers to quality, productivity, and innovation.* Jossey-Bass.

Schulman, P. (1999). Applying learned optimism to increase sales productivity. *Journal of Personal Selling & Sales Management, 19*(1), 31–37. https://doi.org/10.1080/08853134.1999.10754157

Sekerka, L. E., & Fredrickson, B. L. (2008). Establishing positive emotional climates to advance organizational transformation. In N. M. Ashkanasy & C. L. Cooper (Eds.), *New horizons in management. Research companion to emotion in organizations* (pp. 531–545). Edward Elgar Publishing.

Seligman, M. E. (2006). *Learned optimism: How to change your mind and your Life.* Vintage.

Senge, P. M. (1995). *Learning organizations.* Gilmour Drummond Publishing.

Senge, P. M. (2014). *The fifth discipline fieldbook: Strategies and tools for building a learning organization.* Crown Business.

Sharot, T., Riccardi, A. M., Raio, C. M., & Phelps, E. A. (2007). Neural mechanisms mediating optimism bias. *Nature, 450*(7166), 102. https://doi.org/10.1038/nature06280

Sheldon, K. M., Turban, D. B., Brown, K. G., Barrick, M. R., & Judge, T. A. (2003). Applying self-determination theory to organizational research. In *Research in personnel and human resources management* (pp. 357–393). Emerald Group Publishing Limited.

Singer, T., Seymour, B., O'Doherty, J., Kaube, H., Dolan, R. J., & Frith, C. D. (2004). Empathy for pain involves the affective but not sensory components of pain. *Science, 303*(5661), 1157–1162. https://doi.org/10.1126/science.1093535

Snodgrass, S. E. (1985). Women's intuition: The effect of subordinate role on interpersonal sensitivity. *Journal of Personality & Social Psychology, 49*(1), 146–155. https://doi.org/10.1037/0022-3514.49.1.146

Stephens, J. P., Heaphy, E., & Dutton, J. E. (2011). High quality connections. In K. S. Cameron & Spreitzer, G. M. (Eds.), *The oxford handbook of positive organizational scholarship* (pp. 385–399). Oxford University Press.

Thomas, K. W. (1976). Conflict and conflict management. In M. D. Dunnette (Ed.), *Handbook of industrial and organizational psychology* (pp. 889–935). Palo Alto.

Thomas, K. W., & Kilmann, R. H. (1974). *Thomas-Kilmann conflict mode instrument.* Xicom Inc.

Waugh, C. E., & Fredrickson, B. L. (2006). Nice to know you: Positive emotions, self-other overlap, and complex understanding in the formation of new relationships. *Journal of Positive Psychology, 1*, 93–106. https://doi.org/10.1080/17439760500510569

Weiskrantz, L. (1956). Behavioral changes associated with ablation of the amygdaloid complex in monkeys. *Journal of Comparative and Physiological Psychology, 49*(4), 381–391. https://doi.org/10.1037/h0088009

Whalen, P. J., Shin, L. M., McInerney, S. C., Fischer, H., Wright, C. I., & Rauch, S. L. (2001). A functional MRI study of human amygdala responses to facial expressions of fear versus anger. *Emotion, 1*(1), 70. https://doi.org/10.1037/1528-3542.1.1.70

Wolff, S. B., Pescosolido, A. T., & Druskat, V. U. (2002). Emotional intelligence as the basis of leadership emergence in self-managing teams. *The Leadership Quarterly, 13*(5), 505–522. https://doi.org/10.1016/S1048-9843(02)00141-8

Worline, M., & Dutton, J. (2017). How leaders shape compassion processes in organizations. In E. Seppala, J. Doty, M. Worline, D. Cameron, S. Brown, & E. Simon-Thomas (Eds.), *Oxford handbook of compassionate science*. Oxford University Press.

Zaki, J., & Ochsner, K. N. (2012). The neuroscience of empathy: Progress, pitfalls, and promise. *Nature Neuroscience, 15*(5), 675–680. https://doi.org/10.1038/nn.3085

Zhou, J., & George, J. M. (2003). Awakening employee creativity: The role of leader emotional intelligence. *The Leadership Quarterly, 14*(4–5), 545–568. https://doi.org/10.1016/S1048-9843(03)00051-1

Chapter 11

Leadership With Emotional Intelligence Exercises

Exercise 1

Case: You have a subordinate who has low levels of technical and behavioral competencies and presents a low level of self-awareness and acceptance of it, in addition to a low level of accountability, i.e., has difficulties in taking responsibility for their mistakes and correcting them. How should you lead a subordinate with this behavioral profile? What is the best leadership style to adopt? What are the competencies of leadership with emotional and social intelligence most suitable to this situation?

Exercise 2

Case: Your team needs a little more knowledge of internal organizational awareness and about the surrounding market, and the team needs to quickly acquire a set of technical competencies that are essential to achieve the organizational results your company needs. How do you lead a team with this profile? What is the best leadership style to adopt? What competencies of leadership with emotional and social intelligence are most suitable to positively impact this situation?

Exercise 3

Case: Your leadership style has been very demanding from a technical point of view, and the level of technical performance required of your team has been very hard. It is apparent that the team is exhausted, physically and emotionally. Team members begin to show physical and emotional signs of burnout, and you do not know how to act. How do you lead a team in this context? What is the best suitable leadership style to adopt? What competencies of leadership with emotional and social intelligence are most suitable for this situation?

Exercise 4

Case: Your organization will undergo certification of a new management system for processes that will greatly help the organization to optimize and improve the quality of the processes inside the organization. People feel a little lost in the face of this change and the adoption of new processes and procedures. What is the best leadership style to adopt? What competencies of leadership with emotional and social intelligence are most suitable to this situation?

Exercise 5

Case: Your organization will undergo a process of change in structure, with the aim to divide the commercial area by different market sectors. This will initially imply a large change in the organizational structure but also will involve the reengineering of all processes and procedures of commercial service flows. You are the commercial director. What is the best leadership style to adopt in this situation? What are the competencies of leadership with emotional and social intelligence most suitable for this situation?

Exercise 6

Case: You are in an organization where the team and the sector you are responsible for were accustomed to a paternalism from the previous leader, typical of public organizations, where people are afraid of being evaluated for performance, with too much concern for the well-being of the people without accountability for technical, managerial and behavioral competence. This has compromised technical performance and organizational innovation and results. People are quite afraid of performance evaluation and new working methods. They feel threatened and afraid of being fired but have a low level of technical performance and a low level of behavioral competencies, characterized by a strong resistance to change and strong accommodation. The organization has financial stability, as it is public. However, you were hired to bring innovation and positive change to the company. What is the best suitable style of leadership to adopt in this situation? What competencies of leadership with emotional and social intelligence are most suitable to promote innovation and better results for the organization?

Exercise 7

Case: You took over the leadership of a team that is exhausted by the previous bad leader. People were tired of authoritarianism and so much pressure in the technical work and are disengaged. They even have a high level of cynicism and mistrust toward the previous leader and the entire organization. What is the most suitable leadership style to adopt in this situation? What competencies of leadership with emotional and social intelligence are most suitable to deal with this team and this organization?

Exercise 8

Case: You have a very mature team from a technical point of view, composed of some technical high performers. However, the whole team and even some technical high performers have serious ethical and behavioral competence deficiencies. What is the most suitable leadership style in this situation? How do you lead a team like this? What competencies of leadership with emotional and social intelligence are most required to lead this team?

Exercise 9

Case: The relationship with your team is great, but the sector you lead has serious management deficiencies because of other company sectors, which you have no formal power over. What is the most suitable leadership style to adopt in this situation? What are the competencies of leadership with emotional and social intelligence most suitable to helping to solve this problem?

Exercise 10

Case: You disagree with the procedures imposed by your boss to the implementation of the work because you know better technical methods and other ways to do the job. You intend to explain to him this situation, but you are afraid of his negative reaction, to feel questioned in competence. How do you lead in this situation? What is the best leadership style to adopt in this situation? What competencies of leadership with emotional and social intelligence are most suitable to deal effectively and constructively with this situation?

Chapter 12

Answer Suggestions for the Leadership With Emotional Intelligence Exercises

Suggestions for answers are in no way intended to be rigid or considered "unique" or "right," but they are only suggestions based on the leadership model with emotional intelligence and social and leadership styles with emotional and social intelligence and based on 16 years of experience dealing with hundreds of leaders with similar situations discussed in executive education courses.

To evaluate the leadership process, it is very important to consider the analysis of three major variables that interact with each other (Hollander, 1978): the leadership style of the leader, the features of the situation or context where the leadership process is an integral part (Kellerman, 2018), and the technical and behavioral profiles of followers (Kellerman, 2007). These situations were provided by participants from public and private organizations throughout leadership courses with emotional and social intelligence based on their experience and largely discussed in executive education classes in at least six different countries.

Response Suggestion to Exercise 1

According to the leadership model with emotional and social intelligence, one of the fundamental principles of leadership is the ability of the leader to adjust their leadership style at the maturity level of each follower, which is called adaptability. Thus, the principle is to give autonomy and power to followers gradually in the proportion of their technical and behavioral competence development, which means, in practical terms, to give more autonomy and decision power for more mature followers and less autonomy and decision power for less mature ones.

In addition, another principle of leadership is to create a long-term developmental perspective for each follower, from the point of maturity where the follower is to a high-performance point. This perspective can be called growth mindset (Dweck, 2016) applied to leadership for followers. If the follower has a low level of technical and behavioral maturity, and if that adds up to the fact that the follower has a low level of accountability (ability

DOI: 10.4324/9781003503880-15

to take responsibility for their actions and performance), then first, the goal is to increase the level of emotional self-awareness of the follower for the urgent need to take responsibility for their actual performance.

The best suited leadership style with emotional and social intelligence is coercive, to raise the level of emotional self-awareness to make the follower accept the importance of accountability is to use questioning techniques. The idea is to use the questioning technique to induce the follower to take responsibility for their low level of technical and behavioral performance as well as their low level of accountability. The most appropriate style is, in the beginning of the leadership process, the coercive one, until the time when the follower genuinely assumes the responsibility of their technical and behavioral performance level. When the follower genuinely assumes responsibility for their level of technical and behavioral performance, it is critical that the leader change the coercive style to a coach/mentor style.

The leader should prepare a performance feedback interview (where they will use the performance feedback in a coercive style to induce the follower to self-accountability) and then change to the coach/mentor style to present and explain the technical and behavioral developmental plan for the follower, preferably every week, to accompany their technical and behavioral development.

For this follow-up process, the coach/mentor style is the most appropriate because the follower will be able to receive regular systematic feedback about their technical and behavioral competence performance and will gradually develop their competencies.

It is suggested that this developmental plan should take at least 6 months and that the regular feedback should be weekly, with a duration of 20 to 30 minutes.

To use the coercive style adequately, the critical competencies for leaders with leadership with emotional and social intelligence are emotional self-control when encountering natural resistance from the follower and, in turn, with resistance to accountability; the strategic decrease of empathy; and above all, the achievement orientation, because the focus of the leader should be on the task and not on the relationship with the follower.

In the use of the coach/mentor style, the critical leadership competencies are empathy (specifically empathic concern, to stimulate the follower's performance focused on the task), coaching/mentoring and inspirational leadership, so that the follower feels motivated to develop their own technical and behavioral performance, with a growth mindset (Dweck, 2016).

Response Suggestion to Exercise 2

At first, the visionary or authoritative style seems to be the most appropriate for this situation. It is necessary to communicate effectively to the teams and the organization the real situation of the market, the vision for the future of the organization, the real challenges the organization is facing and the real level of performance that is expected from the teams to face these challenges and clearly communicate the level of performance that is expected of each function and how to achieve it.

This message needs to be clear, concise and with high impact. And it needs to be regularly reinforced. It must send a sincere and realistic message. And it is critical to clarify the level of performance that is expected from the team and each follower. This is crucial, because without a vision based on reality, with a clear message about what is expected by each professional from the organization, the organization will hardly achieve its goals.

After the visionary style, it is essential to adopt the pacesetting style, particularly toward high performers and medium performers, if the leader really wants an accelerated acquisition of technical and behavioral competencies that are crucial to the survival of the organization. For lower performing professionals, the coach/mentor style is the most suitable, so that they are regularly monitoring the evolution of their technical performance. If, eventually, in a worst-case scenario, the survival of the organization is compromised by low-performing professionals, it is important to recruit and select the best-performing professionals for each function. If it is a public organization and the organization's policy does not believe that it should be an initiative to be considered, then a coaching and mentoring program is suggested for the lowest-performing professionals to accelerate the low performers toward a middle performance level.

What is known is that, more than ever in the past, the best organizations are gradually starting to have a lower level of tolerance toward low performers, and that is what makes them the best organizations in the market. But all of this must be weighed with great social responsibility on the one hand and the organizational goals to be achieved on the other.

Response Suggestion to Exercise 3

After a context in which the followers suffered the emotional pressure through the overuse of the pacesetting style, the use of the affiliative style seems to be the most suitable style to deal with this situation, to nurture cohesion and well-being of the team again. The affiliative style is very important to help create bonds of interaction and mutual assistance in situations of stress, crisis and emotional or physical exhaustion or burnout (Freudenberger, 1974).

Thus, the leader should invest more in the positive and constructive relationship, particularly with those who find themselves physically and emotionally exhausted. After recovery of this exhaustion, it is important to hear their clear and honest opinions and adopt the democratic/participatory style, so that they have an idea of how the leadership style can help them, so that they can actively participate in the leadership process (Hollander, 1992).

The affiliative style provides autonomy at the level of the relationship, and the democratic/participatory style provides autonomy at the level of organizational decision-making processes, which lets followers participate in decisions about how to direct the work of the team, and this will engage them in an intrinsic and sustainable way in the long term.

Response Suggestion to Exercise 4

The choice of the leadership style with emotional and social intelligence depends on how urgent it is for the organization to complete the certification process. If there is no urgency, the most appropriate style to adopt is the visionary/authoritative style at first, communicating clearly and systematically reinforcing the vision of the need for new processes and procedures for the organization – what is desired for the organization for the next few years, and what is expected of each member of the organization in terms of performance in the future in relation to the implementation of the new management processes.

After the visionary style, because there is no urgency in the implementation process, the coach/mentor style is suggested to systematically monitor followers in learning the implementation processes.

However, if there is urgency in the implementation process, the visionary/authoritative style is suggested, for the same reasons set out above. After the adoption of this style, the most appropriate style is pacesetting, since the organization has a short time frame for learning and assimilating new processes, and the focus should be entirely on the task and the technical performance of the organization, not on people.

After the certification process is complete, change to the affiliative style is suggested to alleviate emotional pressure in professionals in the face of the certification process and to strengthen the trust, affiliation, cohesion and interaction between all members of the organization.

The use of this style is of particular importance not only for the recovery of positive emotions among the team members who feel exhausted but also for the recovery of the psychological vitality and physical and emotional stress of the team to avoid burnout.

Previous scientific studies on psychological vitality point to the importance of positive emotions in the recovery of psychological vitality and intrinsic motivation after a period of great pressure, such as the fulfillment of a demanding technical task within a tight deadline, which is generally considered a factor that tends to decrease intrinsic motivation at work (Amabile et al., 1976) and to generate burnout (Freudenberger, 1974).

Response Suggestion to Exercise 5

In this situation, it is suggested, at first, to adopt the visionary/authoritative style, for the same reasons set out in Exercise 4. After that, the democratic/

participatory style is suggested, with the aim to invite followers to participate in the decision-making processes for the change implementation plan.

One of the classic mistakes of change processes in organizations is the fact that the leaders tend to design the change plan implementation project without obtaining important information about the reality of the organization on the ground, where the strategy of change will be implemented, particularly from teams that work on the ground, in operational levels (McGregor, 1960).

Thus, it is common for a strategic plan or a change implementation plan to have serious implementation problems that could be avoided a priori if it were considered a previous phase between planning and implementation, called organizational implementation adjustment.

In this way, planning should not be imposed to intermediate managers through coercive style toward top managers, and it is suggested that top managers use the democratic/participatory style to involve middle managers in the decision-making processes of change planning before the implementation, so that these intermediate leaders can provide valuable input on the implementation feasibility of the change plan, based on their experience on the ground. This saves the organization from several execution and implementation errors, once the professionals who are on the ground are the most qualified to know whether a measure has a feasibility of implementation or not, allowing a better organizational adjustment process (McGregor, 1960).

It is therefore strategic and emotionally intelligent for top leaders to adopt the democratic/participatory style, so that the intermediate leaders can participate in the processes before their implementation, where they can provide valuable information inputs based on their practical experience, which will allow a better organizational adjustment, between the design process and the implementation process of change.

This organizational adjustment process can result in two scenarios: either the organization acquires more organizational resources to implement a designed aspect of the plan, or the designed aspect of the plan is abandoned if it is not feasible for the organization to acquire new resources.

These are decisions that must be made by top leaders in conjunction with middle-level managers (who are the intermediate leaders of the organization) through the democratic style, where followers who have experience on the ground can save the organization from many implementation mistakes. This style of leadership with emotional and social intelligence, in addition to promoting a real benefit to the organization in the process of change, promotes the intrinsic motivation of the middle-level leaders and the operational team members through the empowerment given to the intermediate leaders (Hollander, 1992). In this respect, this leadership style has two extremely important benefits for the organizational members – one technical and another motivational.

The reality of Portuguese-speaking organizations (and this includes countries such as Brazil, Portugal, Angola, and Cape Verde, among others) is that top leaders tend to use the coercive style in most organizational change processes and in the construction and implementation of the strategic plan. If they often opted for the democratic style, there would be the benefit of improving the organization's capacity to adjust implementation processes, on the one hand, and promoting less resistance to change and greater intrinsic motivation in middle-level leaders and operational team members, on the whole process of change, on the other.

Response Suggestion to Exercise 6

The suggestion is to begin using the affiliative style, with the goal of building positive, trustful, and genuine relationships with people. No one is persuaded by a leader when the leader perceives the relationship as negative or based on fear or authoritarianism.

After the leader was able to build a genuinely positive relationships with people and this is understood by his followers as trustworthy, it is important to change to a visionary/authoritative style. The leader should explain that the situation of poor performance, resistance to change and fear of performance evaluation is not compatible with what is expected of 21st-century organizations, and there is an urgent need to change to an innovative, high-profile vision of performance.

This visionary/authoritative leadership needs to be clear, direct and reinforced regularly until it becomes internalized in the attitudes of the team members, and after its use, there are two possible suggestions. If the organizational change need is urgent, it is essential to apply the pacesetting style, with a focus on tasks and not on relationships, and people will effectively have to adapt to the organization's new requirements and standards of performance.

On the other hand, if the need for change is not urgent, it is preferable to adopt the coach/mentor style, accompanied by a developmental plan of followers for each role. The coach/mentor style allows a process of systematic adaptation and professional growth over time, promotes followers's intrinsic motivation, increases the performance of followers relatively rapidly and systematically, and does not have as many negative effects compared to the pacesetting style.

Thus, if possible, it is strongly suggested to adopt this style to help followers to systematically adapt to new competencies and to be encouraged by the leader to achieve these new levels of performance.

Response Suggestion to Exercise 7

The affiliative style is essential for the leader to give a priority focus on building new relationships of trust. The regaining of confidence in this case is the critical factor, and the leader should explain the context in a clear and transparent way to all followers, to open a space for the construction of trust and a new system of relations between the leader and the followers, based on openness, trust and candor.

After applying the affiliative style, the most appropriate leadership style is the democratic style, and the leader should ask all followers how they would like the leadership to be (based on leader's and follower's behaviors and standards of behavior), to enable the construction of trustful relationships between the leader and the team.

The democratic style is very engaging, particularly after an experience of authoritarianism from the former team leader, and to empower followers to express themselves emotionally in relation to previous situations experienced and to build together with the leader guidelines in which the new leadership practice will occur.

As they feel empowered to express their ideas and feelings and can give suggestions on how they would like to be led, they will feel progressively valued, and confidence and trust are gradually being built (Pugh, 2002). Throughout this process, the transparency and ethics of the leader are very important so that the leader can be trusted by the team (Walumbwa et al., 2008).

Response Suggestion to Exercise 8

This is a very common situation in thousands of organizations worldwide. There is a phenomenon identified by Albert Bandura, professor emeritus of psychology at Stanford University, called moral disengagement (Bandura, 1999, 2004; Barsky, 2011). Moral disengagement consists of psychosocial mechanisms that discourage people from maintaining work ethics within organizations and social groups (Bandura, 2004; Barsky, 2011).

There are several cases reported in the scientific literature of the phenomenon of moral disengagement in organizations (White et al., 2009; Barsky, 2011).

One of these mechanisms is that some leaders, when they are high technical performers, tend to find unethical behaviors more excusable or forgivable, and the ethical aspect is devalued by the whole organization, including subordinates. It is very common for "star performers" in organizations, or professionals with high formal power in organizations, such as directors and CEOs, to be "forgiven" for low levels of behavioral competencies or unethical behavior, when ethical behavior constitutes the most essential part of behavioral competencies.

Thus, it is possible for a professional to be a high performer on technical and managerial competencies, and have very low performance in behavioral competencies due to a lack of ethical behavior, and the organization and/or the social environment where it is part of tend to devalue the behavioral aspect compared to the technical and managerial competencies. There are famous cases of business leaders for whom the media tended to devalue their ethical flaws and describe them as "eccentric behaviors." This phenomenon is what Professor Bandura called moral disengagement (Bandura, 1999).

The seven main mechanisms of moral disengagement are as follows:

Moral justification – It is a "sophisticated" explanation for immoral actions, based on persuasive and well-justified language. A historical example is genocide or terrorist attacks based on religious justifications or ideologies.
Euphemistic language – This consists of the use of language that diminishes the negative value of immoral or unethical behavior. Thus, in an act of corruption, the language "return of a favor," rather than robbery, for example, is used.
The advantageous comparison – This is a comparison made based on an assessment of superiority of some attribute of a group or individual – for example, leaders of an organization make an advantageous comparison against competition; the ideological perception of the superiority of a race can serve as a justification for a genocide; and the perception of superiority over someone from a lower social status or with less power, such as the aggressiveness of a doctor toward a nurse or a CEO toward a cleaning officer, or a PhD student before a recognized full professor.
The diffusion of responsibility – This concerns the fact that everyone is responsible, but the responsibility ends up not falling on anyone, and therefore, there is no one to take over. An example of this is environmental crimes, where everyone is responsible, governments, ministries, businesses, and consumers, but the biggest polluters are those who do not take responsibility for their actions and do not take corrective or preventive measures (Bandura, 2007).
Clearly economic interests and the absence of penalty and prevention make room for dissemination of responsibilities. Only the great leaders assume responsibility for the seriousness of the situation and act at the same time as real needs for change for the crimes being committed. There is, therefore, a clear phenomenon of dissemination of responsibility and moral disengagement of environmental issues at a global level (Bandura, 2007).
The devaluation of the consequences – It may go through discrediting the information, for example, in the case of environmental damage, in relation to scientific data, like the famous statements of former U.S. president Donald Trump, on the environment: "It's very cold out there where the damn is global warming?" (Trump Twitter Archive, 2012).
The attribution of blame – The strategy is to blame others to reduce the environmental responsibility, as can be seen in this quote also from Donald Trump: "The concept of global warming was created by the Chinese and to the Chinese, to make American industrial production uncompetitive" (Trump Twitter Archive, 2012).
Dehumanization – This means removing people's human qualities, and it is the last of the moral disengagement mechanisms. For example, calling a

person a "target" that one will kill in the context of war or calling an enemy "savage" or "barbarian" are dehumanization mechanisms that facilitate moral disengagement and devaluation of an unethical act (Bandura, 1999).

At the level of leadership in an organization, it is up to the leader to take the lead with proactive morality (Bandura, 2004), the responsibility of the situation and to hold each part responsible, including themselves, and initiate an accountability process of all parts involved. The most appropriate style in this situation is the coercive style to provoke an engagement process and to increase awareness of the seriousness of the problem, at the individual and collective level, through the importance of and responsibility for action and its consequences.

The coercive style allows the emotional shock of moral self-awareness engagement and moral self-regulation through accountability of all those involved and paves the way for all to "feel" that they will be responsible if they carry out these unethical behaviors and that it will no longer be acceptable by the organization.

After this style, the visionary/authoritative style is also suggested, so that all these questions are openly discussed, and a new vision of the situation will be created – a vision for which these negative and unethical behaviors will no longer be accepted by the organization, and it is expected that no one will commit them. In addition to promoting a new vision, it is necessary to activate the self-regulation of individual and collective behavior of moral engagement (Bandura, 2004). If the moral engagement process is to be initiated by the leader, it is essential to use the coercive style, to make it clear that unethical behavior is not acceptable and is punishable by clear and unambiguous measures.

But the leader should not remain indefinitely in the coercive style and should change to the visionary/authoritative style to, on the one hand, explain the mechanisms of moral disengagement and, on the other hand, stimulate moral engagement, through the appeal of cognitive empathy and a new positive and ethical view that is expected of all (Bandura, 2004).

Regarding the leader, it is important in this process to appeal to cognitive empathy, from the perspective of positive and constructive values, and to experience this on a real daily basis to build inspirational leadership. More than words, people trust leaders who are consistent between what they say and what they do.

This ethical and moral leadership behavior from the leader will help followers to self-regulate their moral engagement, internalizing the new ethical norms and morality in their senses of "self." There will always be a tendency to find two profiles of followers: those who will internalize genuinely the new ethical rules, intrinsically compromised to the new behaviors and values, with genuine moral engagement – that is, intrinsic morality (Ryan & Connell,

1989); and those who will follow the new ethical rules out of fear of punishment if they do not comply with them: extrinsic, non-genuine moral engagement – that is, extrinsic morality (Ryan & Connell, 1989). The acquisition of morality involves two cognitive mechanisms: the internalization of ethical, prosocial and positive behavior, and self-regulation of unethical behavior (Bandura, 2004). Thus, the leader is suggested to exercise a proactive ethical and prosocial leadership through coercive style and visionary style with emotional and social intelligence (Goleman et al., 2001).

Response Suggestion to Exercise 9

Usually, there is a tendency to think of leadership as only a phenomenon of relationship between the leadership and the subordinates, which is called downward leadership. But leadership is an informal relational process that occurs in 360 degrees, regardless of the formal position where the leader is part. Therefore, leadership can be descending, lateral or ascending.

The most studied and classic case of leadership is descending leadership, which occurs when leaders possess formal leadership positions within organizations and establish descending leadership relationships (which means they occur from top to bottom in the organization hierarchy) with their formal subordinates. But the tendency of organizations is to decrease hierarchical levels within its structure, and the more this occurs, the more that lateral and ascending leadership competencies are needed.

Lateral leadership is a relational phenomenon that occurs among professionals of the same hierarchical level, while ascending leadership is a relational process that occurs when the leader must convince and persuade their own boss about something important for the organization.

In this case, there is the phenomenon of the professional "leading" the boss. The more technical and behavioral maturity from the leader, the more often this phenomenon occurs.

Regarding lateral leadership, it is important that leaders strategically use the visionary/authoritative style to help their peers from other sectors build a common vision of interdependence and cooperation. To this end, the visionary/authoritative style requires the competencies of influence and persuasion, the stimulation of teamwork and collaboration, conflict management and the negotiation of management resources, processes, interdependent tasks and people involved in these areas. It is common to have a silo vision of the organization, and many people have difficulties embracing a systemic and global vision for the organization that intersects the various departmental areas of the organization and its interdependent processes. In industry, conflicts between the commercial department and the production and logistics department are common as well as conflicts in the banking area between the commercial department and the operational department.

In this context, the leader will stand out in a sector if the leader can build positive and productive relationships between the various sectors and is able to persuade and constructively influence their peers to have a more systemic and interdependent view of interdepartmental processes and tasks.

The affiliative style is also important to build relationships based on trust, which will be a valuable resource for the work of interdepartmental teams. As different departments tend to have a limited vision and a "silo mindset" of the organization, the affiliative style between different departments is strategic for the construction of good relationships of trust and cooperation between the various team members of different departments, which will certainly facilitate integrative work between the various sectors.

One of the initiatives that the emotionally intelligent leader can have is the regular practice of meetings and workshops for the analysis of interdepartmental processes, technically called global process focus organizations, including all actors from different areas to analyze organizational processes, especially with a final focus on the client and not on the processes themselves (Gulati, 2009; Porter & Lee, 2013; Drucker, 1993). These workshops should involve managers from different departments to stimulate interdepartmental cooperation and integrative negotiation competencies and high-quality connections between interdepartmental team leaders, and competencies such as influence, persuasion, negotiation, conflict management and teamwork and cooperation (Gulati, 2009).

Still in relation to the specific situation, ascending leadership is also very important, because in addition to convincing their peers from other sectors to carefully analyze interdepartmental management processes, the leader should be able to convince their boss of the relevance of the formal stimulus of its leadership in the work of interdepartmental cooperation.

In this context, the competence of influence and persuasion to the boss becomes a key success factor for the organization in the ascending leadership process. This is the reason why Professor Richard Boyatzis has for decades mentioned the importance of influence and adaptability as critical competencies for management in the 21st century (Boyatzis, 1982), and as one of the most important competencies of leadership with emotional and social intelligence (Boyatzis et al., 2017).

Response Suggestion to Exercise 10

How you position yourself in this situation will depend on the leadership competencies of your boss. To this end, it is very important for the exercise of ascending leadership for the leader to possess two fundamental competencies: cognitive empathy and organizational awareness. Regarding the former, it is essential that the leader understand the boss's perspective, whether the boss has feedback from suggestions for improving the sector, what kind of leadership the boss exercises, whether the boss allows autonomy since you are his formal subordinate, and if the boss has the psychological maturity and flexibility to accept new ideas without feeling their competence questioned.

All this analysis of your boss's behavioral profile should be carefully reviewed and weighed. If the leader feels that the suggestion of improvement can be welcomed in a positive and constructive way, this is the first option suggested in this case. However, if the leader feels that their boss is afraid of being questioned, if the boss possesses an autocratic leadership style in which they do not give space to their subordinates to suggest or propose improvements at work, it is not appropriate to proceed with an improvement proposal, because it will simply not produce constructive or positive effects.

In this situation, the competence of organizational awareness is also important, which means to understand which formal and informal networks of power are at stake when you propose new organizational changes. Identification of the boss's profile and use of competencies such as influence and persuasion are key leadership competencies with emotional and social intelligence to lead in this situation.

If there is a potential for positive influence and persuasion to convince the boss to adopt a better method, and if the organization values the input for organizational improvement in this formal and informal power network, it is strongly suggested that the leader do so.

On the other hand, if the boss's profile is unfavorable to this persuasion process, or if the organization does not value the practice of change (i.e., many organizations have the theoretical discourse that they value innovation, but in practice, when improvements are suggested, this may not be well viewed, so each case should be thoroughly examined), the position to be taken will depend on the importance of change, its urgency and priority, and the consequences for your motivation and engagement as a subordinate.

If this situation is not sufficiently central to the exercise of the work, then not talking to the boss may be the most sensible decision. But if, on the other hand, the subordinate considers that this change is very important for the organization, and if in fact the boss does not take it into account, this could seriously compromise their motivation for work as a subordinate. It is then suggested that the subordinate look for another sector or another organization where they may feel that leadership is open to suggestions and supports autonomy to useful suggestions.

Douglas McGregor, in his famous book *The Human Side of Enterprise* (McGregor, 1960), stated that the potential development of an organization depends on the brain limits of their leaders. If your leaders are limited, you need a cost-benefit calculation that justifies staying in the organization or not or going looking for an organization with which the subordinate identifies the most. Barbara Kellerman, a Harvard Kennedy School leadership professor, states in her book *Bad Leadership: What It Is, How It Happens, Why It Matters* (Kellerman, 2004) that there are seven types of bad leaders: the incompetent;

the rigid; the uncontrolled (without emotional self-control); the one that despises the interests, motivations and needs of its subordinates; the corrupt or unethical; the selfish; and the evil or wicked.

It is the responsibility of the subordinate, in the event of a situation of bad leadership, if that is the case, to proceed to an analysis of your leader's profile and calculate the cost-benefit of your behavior position: whether you should obey, try to persuade the boss, or, at the limit, leave the organization. Exiting an organization with which a subordinate does not identify with constitutes in this case an act of upward leadership from the leader.

In the evolution of the scientific study of organizational commitment – which is the factor that causes professionals to remain committed to an organization – Meyer and Allen (1991) found that, increasingly, professionals are not so much linked to an organization for affective reasons, for normative factors or for continuity factors, but the trend of organizational behavior of professionals in the 21st century will be to develop a link to the profession they practice and not to organizations (Meyer & Gagne, 2008).

For example, if an organization provides all the technical, financial, behavioral and environmental issues for the development of the potential of the professional, then the professional will establish a commitment to the organization. However, if the employee feels that staying in an organization will jeopardize the development of their potential in the long term, then the level of organizational commitment will be low, and the employee will look for another organization that can help them develop their professional potential.

This means that the commitment will tend to be toward professional excellence and not toward the organization (Meyer & Gagne, 2008). Today, most organizations charge a "loyalty" (the scientific term is "organizational commitment" as previously mentioned) of the employee to the organization, but often the organization and its leaders do not realize that they do not give the necessary elements for the development of the professional potential of their subordinates, that is, to meet their professional developmental needs (Kellerman, 2004).

Therefore, most leaders who receive a request for resignation from a leader tend to face the employee in a negative and hostile way, as if the employee was being "ungrateful," "unfaithful" or was "betraying" the organization, without realizing that it was the organization and its leaders who have not created, through the lack of leadership actions, the conditions for development of the professional activity of the subordinate. We could call this reaction from the leader of the organization "bad leadership without self-awareness" (Kellerman, 2004).

Usually, the implementation of the process of exiting an organization is a slow process in the mind of the subordinate, in which the tendency of dissatisfaction toward the organization and their leadership has been happening for a long time. The emergence of better professional alternatives in the market is what constitutes the motivational engine of change, or when the deterioration of the subordinate's relationship with the organization becomes unbearable, or even the existence of some conflict, which tends to speed up the exit process. In any case, the subordinate plays the role of calmly pondering the cost-benefit of an exit and seeking a better alternative for their professional development, following their commitment to the profession, a term that can be designated as "professional excellence commitment."

Commitment in this case is to professional excellence inside a specific profession, regardless of the organization where the employee is working. Thus, increasingly, organizations will only be a means through which the employee can develop their potential (Meyer & Gagne, 2008). If, for some reason, it sees its development negatively compromised by the organization, it is even desirable that this professional look for another organization that is more identified with the professional competencies and the development of their potential.

And this is a leadership position from the part of subordinates. Many professionals block their developmental potential by remaining in an organization that does not stimulate its growth, only by a logic of continuity and financial security, which seriously compromise the intrinsic motivation of long-term professionals and produce professional mediocrity, dissatisfaction and disengagement.

Organizations that provide the conditions for professional development and excellence to their employees, to ensure the satisfaction of three basic needs for intrinsic motivation (i.e., autonomy, competence, attachment), as well as meet the needs of autonomy for professional development, provide the development of technical and behavioral skills and a positive and constructive environment, will be the organizations best prepared to retain the talent (Meyer & Gagne, 2008).

References

Amabile, T. M., DeJong, W., & Lepper, M. R. (1976). Effects of externally imposed deadlines on subsequent intrinsic motivation. *Journal of Personality and Social Psychology, 34*(1), 92.

Bandura, A. (1999). Moral disengagement in the perpetration of inhumanities. *Personality and Social Psychology Review, 3*(3), 193–209. https://doi.org/10.1207/s15327957pspr0303_3

Bandura, A. (2004). Selective exercise of moral agency. In T. A. Thorkildsen & H. J. Walberg (Eds.), *Nurturing morality. Issues in children's and families' Lives* (vol 5). Springer.

Bandura, A. (2007). Impeding ecological sustainability through selective moral disengagement. *International Journal of Innovation and Sustainable Development, 2*(1), 8–35. https://doi.org/10.1504/ijisd.2007.016056

Barsky, A. (2011). Investigating the effects of moral disengagement and participation on unethical work behavior. *Journal of Business Ethics, 104*(1), 59. https://doi.org/10.1007/s10551-011-0889-7

Boyatzis, R. E. (1982). *The competent manager: A model for effective performance*. John Wiley & Sons.

Boyatzis, R., Guise, S., Hezlett, Kerr, P., & Lams, S. (2017). *Emotional and social competency inventory research guide and technical manual*. Korn Ferry Technical Manuals.

Drucker, P. (1993). *Post-capitalist society*. HarperCollins.

Dweck, C. (2016). What having a "growth mindset" actually means. *Harvard Business Review, 13*, 213–226.

Freudenberger, H. J. (1974). Staff burnout. *Journal of Social Issues, 30*, 159–165. https://doi.org/10.1111/j.1540-4560.1974.tb00706.x

Goleman, D., Boyatzis, R., & McKee, A. (2001). *Primal leadership: Unleashing the power of emotional intelligence*. Harvard Business Review Press.

Gulati, R. (2009). *Reorganize for resilience: Putting customers at the center of your business*. Harvard Business School Press.

Hollander, E. P. (1978). *Leadership dynamics: A practical guide to effective relationships*. Free Press.

Hollander, E. P. (1992). Leadership, followership, self, and others. *The Leadership Quarterly, 3*(1), 43–54. https://doi.org/10.1016/1048-9843(92)90005-Z

Kellerman, B. (2004). *Bad leadership: What it is, how it happens, why it matters*. Harvard Business Press.

Kellerman, B. (2007). What every leader needs to know about followers. *Harvard Business Review, 85*(12), 84–94.

Kellerman, B. (2018). *Professionalizing leadership*. Oxford University Press.

McGregor, D. (1960). *The human side of enterprise*. McGraw-Hill.

Meyer, J. P., & Allen, N. J. (1991). A three-component conceptualization of organizational commitment. *Human Resource Management Review, 1*(1), 61–89. https://doi.org/10.1016/1053-4822(91)90011-Z

Meyer, J. P., & Gagne, M. (2008). Employee engagement from a self-determination theory perspective. *Industrial and Organizational Psychology, 1*(1), 60–62. https://doi.org/10.1111/j.1754-9434.2007.00010.x

Porter, M. E., & Lee, T. H. (2013). The strategy that will fix health care. *Harvard Business Review 91*(10), 50–70.

Pugh, S. D. (2002). Emotional regulation in individuals and dyads: Causes, costs and consequences. In R. G. Lord, R. J. Klimoski, & R. Kanfer (Eds.), *Emotions in the workplace: Understanding the structure and role of emotions in organizational behavior* (pp. 147–182). Jossey-Bass.

Ryan, R. M., & Connell, J. P. (1989). Perceived locus of causality and internalization: Examining reasons for acting in two domains. *Journal of Personality and Social Psychology, 57*(5), 749–761. http://dx.doi.org/10.1037/0022-3514.57.5.749

Trump Twitter Archive (2012). www.trumptwitterarchive.com

Walumbwa, F. O., Avolio, B. J., Gardner, W. L., Wernsing, T. S., & Peterson, S. J. (2008). Authentic leadership: Development and validation of a theory-based measure. *Journal of Management, 34*(1), 89–126. https://doi.org/10.1177/0149206307308913

White, J., Bandura, A., & Bero, L. A. (2009). Moral disengagement in the corporate world. *Accountability in Research, 16*(1), 41–74. https://doi.org/10.1080/08989620802689847

Conclusion

Neuroscience today, and increasingly in the future, plays a revolutionary role not only in emotional intelligence learning but also in all areas of psychology. Any department of psychology in the world that does not have as a scientific method of study neuroscience techniques – fMRI, brain stimulation techniques (e.g., transcranial magnetic stimulation and transcranial direct current stimulation) and the electroencephalogram – will be left behind in the findings not only about the brain but also about psychology, as happened with those departments of the 1960s that insisted on maintaining psychoanalysis and did not change to experimental psychology. So, we can say that neuroscience will change the future of psychology research, and its potential applications are extraordinary.

In this book, we tried to demonstrate how understanding the brain mechanisms of emotions and emotional intelligence can shed light on the phenomenon of emotional intelligence.

But the revolution that neuroscience will provide in psychology is still to come. There are two scientific discoveries in neuroscience with a huge potential for application: "mind reading," using fMRI and real-time fMRI feedback to individuals.

Mind reading was the result of pioneering scientific work developed by the laboratory neuroscientist Marcel Just of Carnegie Mellon University in the United States. The technique of mind reading is done in two phases: the first is prior training of a participant reading a series of words inside the fMRI machine; after training, in a second phase, the same participant is asked (also inside the fMRI) to think about one of the words they read in the first phase.

Functional magnetic resonance imaging can predict what word participants are thinking about, through the prediction of neural activation associated with the word, through data collected in the participant's first reading phase of the experiment (Mitchell et al., 2008).

This mind-reading technique has extraordinary potential in terms of therapy, psychology and education, as it allows operationalization of the neural individualized feedback about what the individual is thinking or feeling in real time, through the mind-reading technique combined with the feedback technique, through real-time fMRI.

This technique changes everything in terms of technology applied to education. This would be a dream come true for Wilhelm Wundt, one of the founding fathers of experimental psychology, with his laboratory in 1848 (Mandler, 2011), once this is a technology that allows you to give feedback about a participant's learning brain (also called neuroplasticity by neuroscientists) to the individual themselves, constituting a learning process (i.e., education) based on gradual brain changes in the individual with specific emotions, emotional self-awareness, emotional self-control and thought processes.

Due to the focus of this work, we do not go into more detail about this upcoming revolution, because our goal here is just to arouse the reader's interest and curiosity for the revolutionary potential that neuroscience will have in the application of its technology and knowledge to education in general and to the education of emotional intelligence in particular.

There is a clear need for a massive investment in emotional intelligence education of children and adolescents, an investment that should begin with the training of teachers and all school staff and parents, with programs of recognized quality, such as those mentioned in this book. The benefits are immense.

Deep learning of emotional intelligence over a long period of time prevents depression and anxiety in a way that significantly promotes levels of happiness and well-being as well as the quality of relationships.

There is also a lack of investment in the education of emotional intelligence of adults, and this is worrying. Economic, COVID-19 pandemic, social and climate crises pose great challenges to humanity, and emotional intelligence can be a strong protector and generator of emotional resilience in the face of global adversities.

Regarding the executive education of emotional intelligence in organizations, according to some specialists from universities such as Harvard University (Kellerman, 2018) and Stanford University (Pfeffer, 2015), most leadership executive education is not effective.

The poor quality of training generated a perception on the part of the leaders of organizations that leadership with emotional intelligence was a "management fashion" of the first decade of the 21st century that has passed. Currently, leadership executive education has "other fashions." According to these experts, leadership training is taught by experts with a superficial and weak background in scientific terms, who are superficial and who are too short in time (Pfeffer, 2015; Kellerman, 2018).

Business schools themselves have done a terrible job concerning this, according to Pfeffer (2011, 2015). To change this situation, there is a need for a better-quality education of teachers and trainers from a scientific point of view, a longer assessment and more focus on accurate measurement of each participants' competencies, with a personal development plan that is well defined over time, with an emphasis on experiential learning (focused on the specific reality of each individual), with a minimum duration of 6 months to 1 year, and its periodic follow-up by a good quality specialist.

These changes would provide to the field of emotional intelligence the professional dignity it deserves.

Emotional intelligence learning is a lifelong learning process. It requires focus on emotional self-awareness in the most diverse situations, self-observation of the many personal patterns of emotions and the generation of positive and constructive emotions connected to intrinsic values cultivated by everyone.

When taken seriously and addressed in a profound way, it can be highly transformative and lead to a genuine evolution of the self.

In this context, the evolution of the self means greater emotional stability in the face of adversity, more wisdom in decision-making and the choice of values for life, and deeper quality relationships with others and with oneself.

References

Kellerman, B. (2018). *Professionalizing leadership*. Oxford University Press.

Mandler, G. (2011). *A history of modern experimental psychology: From James and Wundt to cognitive science*. MIT press.

Mitchell, T. M., Shinkareva, S. V., Carlson, A., Chang, K. M., Malave, V. L., Mason, R. A., & Just, M. A. (2008). Predicting human brain activity associated with the meanings of nouns. *Science, 320*(5880), 1191–1195. https://doi.org/10.1126/science.1152876

Pfeffer, J. (2011). Leadership development in business schools: An agenda for change. In J. Canals (Ed.), *The future of leadership development* (pp. 218–237). Palgrave Macmillan.

Pfeffer, J. (2015). *Leadership BS. Fixing workplaces and careers one truth at a time*. Harpercollins Publishers.

Index

Note: Page numbers in *italics* indicate a figure and page numbers in **bold** indicate a table on the corresponding page.

accountability 184, 191–192, 211–212
achievement orientation 169, 176
achievement-oriented pride 44
action tendencies, emotions 4, 57, 81, 126
adaptability 169, 173–175, 211
admiration 26–27, 63, 182; neural basis of 27
advantageous comparison 218
affiliative style 184–185, 213, 214, 216–217, 221
aggressiveness 29
Allen, N. J. 223
altruistic punishment 31
amygdala: anger and 7–8; contempt, and 31, *32*; embarrassment and 36; emotional intensity and 84; fear and 5, *5*;
anger 6–7, 57, 174; fear and 4–5; neural bases of 7–8; and contempt 31; as moral emotion 27–29; sadness and 8–9; self-awareness and emotional self-control in situations and 142, **143–144**
anterior cingulate cortex 27, *28*, 83–85, *84*
anterior insular cortex/anterior insula 81–82, 84
anterior subgenual cingulate cortex 21, 94
Aquino, T. 44
Aristotle 33, 43, 44
The Art of Loving (Fromm) 102
authenticity 167, 171, 172
authoritarianism 203–204, 216, 217
authority 179

Bad Leadership: What It Is, How It Happens, Why It Matters (Kellerman) 222–223
Bandura, A. 116, 217
basal ganglia 10–12, *12*, 16
basic emotions: anger 6–8; biological mechanisms 4; characteristics 3; constructive/destructive aspect 4; disgust 9–12; fear 4–5; happiness and joy 13–16; reward system 16–18; sadness 8–9; surprise 12–13, *13–14*
benevolence 19, 22, 23, 138
biology 74, 104–107
blame, attribution of 218
Boyatzis, R. E. 77, 126, 136, 168, 221
broaden and build theory of positive emotions 57–60: broaden hypothesis 60–61; build hypothesis *see* construction hypothesis

cardiovascular recovery 58, 62
central values for self *see* self-centrality
coaching 170, 180–181
coaching style 186, 212, 213, 216
coercive style 183–184
cognitive empathy 105, 118, 138, 177, 219, 221
cognitive reappraisal 6, 88; dorsolateral prefrontal cortex 91–94, *92*, *94*; emotional regulation 91
cognitive regulation strategies: anticipation 90; cognitive reappraisal

91–94; emotional response 89, 90; non-pain stimulus and neural bases 90; top-down and bottom-up responses 90–91; ventral anterior cingulate cortex 94–96, *95*
Collaborative for Academic, Social, and Emotional Learning (CASEL) program: relationship management 118; responsible decision-making 119–120; self-awareness 116; self-management 116–117; social awareness 118; statistical analysis 115
communication 118, 137, 173
compassion 19–21, *20*, 63, 138
complexity 74–77, 101–102
condemning others: anger/moral indignation 27–29; contempt 31, *32*; moral disgust 29–31, *30*
conflict management 170, 181–182, 184
conscientiousness 67
consciousness 85–86, 89
consequences, devaluation of 218
consistency-based influence strategy 179–180
construction hypothesis: AIDS, mortality of 66; anticipation 62; appreciation 61, 62; cognitive resources 61; conscientiousness 67; depression 64–65; disease symptoms of 64; flu symptoms 65–66; groups 67; hypertension, diabetes and respiratory tract infections 66; learned optimism 63; life goals 63; longevity 66; loving-kindness meditation 67; mindfulness 61; parasympathetic nervous system, vagal tone 65; perception 64; positive relationships 64; resilience 62; self-acceptance 62–63; social support 63–64; telomeres 66–67; thinking and decision-making strategies 61
contempt 31, *32*
contentment *see* serenity
coping mechanisms 88–89
Csikszentmihalyi, M. 85

Damasio, A. 5, 21, 25, 27, 35, 59, 81, 110
Darwin, C. 4–6, 27
decision-making strategies 60–61, 177, 183, 185, 215
deep learning 227
default mode network 77
dehumanization 218–219
democratic style 185, 213
depression 64–65
devaluation of consequences 218
diabetes 66
Diener, E. 15
differentiation 74–75, 77, 107
diffusion of responsibility 218
discoveries: definition 126–127; new behaviors, practical experimentation of 134; self-directed learning project 134–137
disgust 9–10, 57; moral 29–30; neural bases of 10–12; neural bases of moral disgust 30–31
diversity, appreciation of 118
dorsal anterior cingulate cortex 39, *39*, 82, *84*, 90–91
dorsolateral prefrontal cortex 46, *46*, 79, 91–94, *92*, *94*
dorsomedial prefrontal cortex 77, *78*, 85, *85*, 90
downward leadership 220
drug addiction 18
Druskat, V. 182
Dutton, J. 181

eccentric behaviors 217–218
education: CASEL program 115–120; children and adolescents 89, 129, 133; emotional intelligence 45, 115, 120–122, 227; executive classes 211; intrinsic and extrinsic self 129; motivation 131; organizations 67, 75, 88; pride-based distortions 45; psychology 226; RULER program 120–122; self 126–138; self-control 29
Einstein, A. 29, 41, 42
embarrassment 34–35; neural basis of 35–36
Emmons, R. A. 23
emotional empathy 118, 138, 177
emotional literacy 84, 95, 121, 125
emotionally intelligent (constructive) management: anger 6; fear 5; sadness 9

emotional regulation: behavioral strategy 89–90; cognitive *see* cognitive regulation strategies; vagal tone 65
emotional self-awareness 80–81, 88, 116, 169, 184, 212; anterior cingulate cortex 83–85, *84*; anterior insula 81–82, *82*; dorsomedial prefrontal cortex 85, *85*; evolution and expansion of 85–86; leadership 170; medial orbitofrontal cortex 82–83, *83*
emotional self-control 46, 80–81, 88–89, 116–117, 169, 212; leadership 170; regulation *see* emotional regulation; relationships 147, **147–150**; situations 142, **142–146**
emotional suppression 88–90
emotion identification 116
emotions: basic *see* basic emotions; characteristics 57; children 121, 122; educational organizations 88; functionality/dysfunction 120; human 22; James-Lange's theory 81; love 107, 109; moral *see* moral emotions; negative *see* negative emotions; neural bases 3; neuroscience 174; positive *see* positive emotions; sadness and neutral stimulus 58; *vs.* sentiment 109–110; suppressive regulation 91
empathic concern 21, 138, 177, 212
empathy 19, 118, 169, 177, 183, 184; positive emotions and 60–61; compassion, benevolence and 138
envy 36–39; neural bases of 39, *39*
euphemistic language 218
euphoria 15–16, 100
evolution 4: importance and expansion of awareness 85–86; of self 74–77, 107–109; self-acceptance and 62–63
exercise: for children 126; extrinsic/intrinsic values of life 157, **157–164**; impulsivity self-control and delay of gratification 150, **151–152**; purpose of life 155, **155–156**; self-awareness and emotional self-control in relationships 147, **147–150**; self-awareness and emotional self-control in situations 142, **143–147**; self-discipline and creation of good habits 153, **153–154**; *see also* leadership exercises
experimental psychology 21, 76, 226, 227

fear 4–5, 57; adaptability and 174–175; conditioning 95; neural basis of 5; emotional self-awareness and self-control exercise 142, **143**, **145–146**
fear-based leadership 174–175
financial stability 201–202
flight-or-fight system 57
Fredrickson, B. 57, 59, 62
Freud, S. 88
Fromm, E. 102, 103
fun 47
functional magnetic resonance imaging (fMRI) 13, *13–14*, 226

Gandhi, M. 29
generalization 176
global process focus organizations 221
Goleman, D. 125, 168, 183
good habits 153, **153–154**
Gouldner, A. W. 22
gratification, delay of 89, 151
gratitude 21–25, *24*, *25*, 41, 62, 86
growth mindset 108, 109, 126, 175, 211, 212
guilt 31–33; neural bases of guilt 35, *35*

Haidt, J. 18
happiness/joy 13–16; gratitude and 23; and health 66; intrinsic vs. goals and 130–131; neural bases of 16
high-quality connections 48, 109, 131, 181
hope 42; neural bases of 42–43, *43*
How Gertrude Teaches Her Children (Pestalozzi) 129
human relationships 100, 104–107, 110; *see also* relational systems; relationships
The Human Side of Enterprise (McGregor) 222
hypertension 66

ideal self 127–132
illusions 15–16, 76, 99–101, 104, 108
impulsivity 17, 110, 151

impulsivity self-control exercise 151, **151–152**
inferior frontal gyrus 78
influence 33, 109, 168, *169*, 170; authority 179; children and 126; consistency 179–180; definition 178; emotional responses 90; liking 178–179; reciprocity 179; scarcity/rarity 179; social interaction 178; social proof 178
inspiration 27, 40, 47
inspirational leadership 170, 182, 184, 219
insula 10, *11*, 27, *28*
integration 74–75, 77
interest 42, 129, 137, 179, 182, 218
interpersonal behaviors 38
intrinsic motivation 40–42, 64, 117, 131–132, 224
Isen, A. 60

jealousy 38; *see also* envy
Jobs, S. 182
joy 13–16, 40–41, 57, 86; *see also* happiness
Just, M. 226

Kahneman, D. 15
Kellerman, B. 222

leadership: competencies 169–170; downward 220; emotional and social intelligence 168, *169*, 172, 211, 212; emotional stress 167; executive education 227; inspirational 182, 184, 219; lateral 220; levels of analysis 168; principle 211; survival and continuity 167; training 227; training programs 167; variables 211
leadership exercises 191–210; answer suggestions 211–224
leadership styles: affiliative 184–185; authoritarianism 203–204; certification, new management system 197–198, 214; coaching 186; coercive 183–184; democratic 185; emotional and social intelligence 183, **186**, 212; internal organizational awareness 193–194; management deficiencies 207–208; market sectors 199–200; pacesetting 185–186; paternalism 201–202; technical and behavioral competencies 191–192; technical high performers 205–206; technical performance 195–196; visionary/authoritative 184; work implementation 209–210
learned optimism 63, 176
liking strategy 178–179
longevity 66
love 48, 107–109
loving-kindness meditation 64, 67

Mandela, N. 182
mania 15, 16
Maslow, A. H. 76
materialistic values 23, 131, 138
Mayer, J. D. 120
McClelland, D. 176
McCullough, M. E. 22, 23
McGregor, D. 174, 222
McKee, A. 168
meaningful relationships 118
medial orbitofrontal cortex (mOFC) 21, 82–83, *83*, 91, *92*
medial prefrontal cortex (MPFC) 9, *10*, 58, 77, *78*
mental beliefs 90–91
mentoring 170, 180–181
mentor style 212, 213, 216
Meyer, J. P. 223
mindfulness 61, 67
mind reading technique 226–227
moral barometer 22, 23
moral behavior of benevolence 22, 23
moral disengagement 119, 171, 182, 217–219
moral disgust 29–31, *30*
moral elevation 25–26, 63, 86, 182; neural bases of 26, *26*
moral emotions: admiration 26–27; anger/moral indignation 27–29; classification 18; compassion 19–21, *20*; condemning others 27–31; contempt 31, *32*; embarrassment 34–36; enhancing others 18, 21–27; envy 36–39, *39*; gratitude 21–25, *24*, *25*; guilt 31–33, 35–36; moral disgust 29–31, *30*; moral elevation 25–26, *26*; Schadenfreude 36, *37*; self-conscious 31–36; shame 33–34, 35–36; toward suffering of others 18, 19–21

moral indignation 27–29; *see also* moral disgust
moral justification 218
moral motivation 22–23, 106
mortality 66
motivational belief 109–110
motivational system 17, 59, 76

negative emotions 3; debt 23; dysfunctional 121; envy 36; events 62; interferences 95; neural and behavioral mechanisms 15; physiological changes 57; probability of survival 57; self-awareness exercise **149–150**; stressful events 59
neuroplasticity 20, 227
neuroscience: biology and human relationships 104–107; cognitive 16; emotions 174; envy 39; fear 5; positive and negative emotions 59; scientific method 226; social 3; wanting concept 17
neutral stimulus 58, 91
non-emotional (destructive) management: anger 7; fear 5

On Becoming a Person (Rogers) 63
orbitofrontal cortex 7, 7–8, 10, *11*
organizations: awareness 117, 169, 177–178, 221, 222; commitment 223; donations 31; envy 37–38; games 40; implementation adjustment 215; neuroscientific research 19; resources 64, 215; skills 88, 117; training 38
others 125, 128; benefit of 86, 155; broaden hypothesis 60–61; leadership with emotional intelligence 168–170; psychological maturity system 101–104; self *vs.* 78, 138; social awareness and 118; social support to 63–64; symbiotic relationship system 99–101; *see also* moral emotions
oxytocin 105–107

pacesetting style 185–186, 213–214
parasympathetic nervous system 65
parenting styles 132
parents communication techniques 137
permanence 176
personal development 109, 126, 130, 131, 133

personalization 176
perspective taking 118
persuasion 178, 222
Pestalozzi, J. H. 129
Pfeffer, J. 178, 227
physical health 64, 65
positive emotions 3; admiration 47; cognitive domains 60; fun 47; gratitude 41; hope 42–43, *43*; inspiration 47; interest 42; joy 40–41; love 48, 108; physical resources 64–65; pride 43–46; psychological resources 62; reassessment 61; recovery 214; relationships 147, **147–148**; right prefrontal cortex 59; serenity 41–42; social resources 63–64; stimulus 58; typology of 31; vagal tone 65
positive outlook 169, 176–177
positive psychology 38, 85
positron emission tomography (PET) 81
post-traumatic growth 9
pregenual anterior cingulate cortex (pACC) 94
pride 43–46; neural basis of 46
Primal Leadership (Goleman, Boyatzis and McKee) 168
professional excellence commitment 223, 224

real self 132–133
reciprocity 22, 23, 101, 179
Recognize, Understand, Labeling/Naming, Expressing, Regulating/Regular (RULER) program 120–122
relational systems: psychological maturity system 101–104; respect 103–104; symbiotic relationship system 99–101
relationships: amygdala 8; conflicts and breakups 7; emotional self-control 147, **147–150**; emotion and sentiment 109–110; emotions and social 105–107; empathic concern 21; empathy, compassion and benevolence 138; interpersonal 9; intrinsic values 104; love 41, 107–109; management 118; meaningful 118; modern 103; psychological maturity 101; quality 32, 44, 46; responsibility 103; self-awareness 147, **147–150**; selfish

values 104; surprise emotion and distributed brain network 13
repression 88, 89
resilience 42, 45, 62
respiratory tract infections 66
responsibility: decision-making 119–120; diffusion of 218; ethical and moral 119; relationship with other 103; social 213; subordinate 223; technical and behavioral performance level 212
resting state 7, 8, 77
reward system 16, 17, 21, 24, 106; joy and pleasure experience 16; liking mechanisms 17; types 16; wanting mechanisms 17–18
Rhetoric (Aristotle) 43
rhinovirus infection 65–66
Rogers, C. 63

sadness 8–9; neural basis of 9
Salovey, P. 120
scarcity/rarity 179
Schadenfreude 36; neural bases of 36, *37*
Schore, A. N. 82
self 125–127; evolution of 74–77; ideal 127–132; ideal vs. real 45; and love 108–109; neural bases of 77–78, *79*; and others 99–104; real 132–133; *see also* discoveries
self-acceptance 62–63, 135
self-awareness 38, 45, 80–85, 116; exercises **142–146, 147–150**
self-centrality 9, 130
self-central values 126
self-confidence 116, 176
self-consciousness: anterior insula 27, 81; embarrassment 34–36; guilt 31–33, *35*; shame 33–34
self-directed learning model 77, 126; construction of 134; mentors 136–137; social environments 134–136
self-discipline 117, 153; exercises **153–154**
self-efficacy 116, 122
self-management 116–117, 172
self-reflection 27, 76, 85, *85*, 137
Seligman, M. 63, 176
Senge, P. 125, 173
sentimental self-control 109–110

serenity 41–42
setting goals 117
shame 33–34; neural bases of 35–36
silo mindset 221
Simmel, G. 22
Singer, T. 177
Smith, A. 21, 22
social and emotional learning (SEL) 115–120
social awareness 118, 168, 172
social dimension 74, 172
social environments 15, 134–136, 217
social influence 178
social involvement, active 118
social proof 178
social support 63–64, 118, 136
Spencer, H. 4
stress management 88, 117
subgenual anterior cingulate cortex 21, 94
suffering of others: admiration 26–27; compassion 19–21, *20*; gratitude 21–25, *24*, *25*; moral elevation 25–26, *26*
suppressive emotional regulation 89–90
surprise 12–13, *13–14*
symbiotic relationship system 99–101

teaching method 129
teamwork 118, 170, 182
telomeres 66, 67
Tesser, A. 23
The Theory of Moral Sentiments (Smith) 21
Thomas, K. 181
The Triple Focus (Goleman and Senge) 125
Trump, D. 218

vagal tone 65, 67
ventral anterior cingulate cortex (vACC) 90, 94–96, *95*
ventromedial prefrontal cortex (vmPFC) 58, 93, *93*
visionary/authoritative style 184, 212–214, 216, 219

de Wall, F. 104
The Wealth of Nations (Smith) 21
Weissberg, R. 115
Wundt, W. 227